UNDERSTANDING & USING MODERN ELECTRONIC SERVICING TEST EQUIPMENT

To Polly—without whom this would not have happened.

No. 777
$8.95

UNDERSTANDING & USING MODERN ELECTRONIC SERVICING TEST EQUIPMENT

BY CHARLES M. GILMORE

TAB BOOKS
Blue Ridge Summit, Pa. 17214

FIRST EDITION

FIRST PRINTING—DECEMBER 1976

Printed in the United States
of America

Hardbound Edition: International Standard Book No. 0-8306-6777-6

Paperbound Edition: International Standard Book No. 0-8306-5777-0

Library of Congress Card Number: 76-45062

Contents

Introduction

There has been a radical change in the sophistication of low-cost instrumentation in the past few years as well as a rather dramatic increase in the complexity of commercial and consumer electronic products. Both of these reasons have contributed to the use of instruments by persons who do not fully understand them. To add to all of this, there seems to be no other source than the instrument manufacturer for determining just what test equipment specifications mean. I hope that this handbook will help correct that particular problem.

This is not a textbook. It is therefore not intended to be read from cover to cover in one or more sittings. It is instead meant to be a reference book; one the user consults whenever a question arises concerning an electronic instrument. To aid in this use, a common format has been chosen for discussing each instrument.

Each of the chapters is devoted to a particular instrument. The chapters open with a brief review of the fundamentals behind the instrument. These discussions are at the system level and are not detailed circuit descriptions. Following the review of fundamentals, there is a detailed discussion of the specifications applied to instruments of this type. This section defines each specification in detail, indicates its relative worth in comparison to the other specifications, and explains the alternate ways used to specify the same characteristic.

Often the features of an instrument are as important as the specifications, so the third section of each chapter delineates features and characteristics of the instrument. I differentiate between specifications and features with the following rule. If a number is applied to a characteristic of an instrument and that number has a tolerance of error, then the number is a *specification*. If the number does not have a tolerance of error (such as the number of digits on a digital frequency counter), then I define this characteristic as a *feature*. Features can make an instrument easy to use, but the lack of proper features can make it difficult to operate the instrument at its full capability.

The last section of each chapter deals with applications of the instrument. There are two areas to be covered here—first, useful applications of an instrument which display the features and specifications discussed earlier and, second, errors which creep into a measurement made by the instrument. Errors are divided into operator errors and instrument errors.

Of course, it would take many books this size to cover all the electronic instruments available, not to mention the nonelectronic instruments such as VOMs, tube testers, bridges, etc. The instruments I have chosen are electronic; they employ active electronic devices. All are available at low cost. Instruments in any category discussed are available with the features and specifications discussed for less than $1000. This does not mean versions of these instruments are not available for a great deal more than $1000, but it does mean that, in many of the categories of instruments discussed (oscilloscopes being a notable example), there are thousand-dollar-plus versions having specifications and features not discussed in this handbook. The instruments discussed are those which make an electrical measurement; they are common and fundamental. While there are many instruments which are very special in application, they are not discussed here. Other instruments are nothing more than a number of fundamental instruments packaged in one box for the convenience of one particular user. These, once recognized, can easily be understood in terms of their fundamental parts and so are not included here. The instruments covered are those of interest to the experimenter or servicer, and not those whose use is almost exclusively confined to the developmental laboratory.

I hope this handbook will help the instrument user in three ways: (1) in determining when a given class of instrument will do a job, (2) in selecting the correct instrument from the large number offered, and (3) in operating the instrument after purchase (especially if the instrument or a particular feature of the instrument has not been used for some time).

To purchase an instrument without a thorough understanding of its specifications and features and how they relate to the characteristics of other available instruments frequently leads to a waste of the instrument dollar. To carry the selection process to the fullest, I strongly recommend the potential purchaser of an instrument construct a matrix with which to compare the various specifications and features on an equal basis.

Make a list of all the specifications and features of an instrument, by name, down one side of a sheet of paper. (The sections of this handbook on specifications can be used to create this list. As an alternative, copy the list of specifications as given by one of the manufacturers under consideration.) This list comprises the *Specifications* column. Each succeeding column will list the published specifications and features of each model to be considered. Don't forget the price—it is usually a prime specification! Once the table has been created, fill in the numbers. Here the fun starts, and the *Specifications* and *Features* sections of this handbook should help. Many of the manufacturers list some of their specifications and features in a slightly different, or altogether different, manner from the others. You have to interpret the exact common specification before entering the numbers on the chart. Once the numbers are filled in, you can compare them against your requirements (another column perhaps?) and select your best buy.

The appendix of this handbook contains front-panel photographs of a number of instruments and their specifications. This is not intended as a catalog, but rather as a place where the reader can compare the actual specifications of an instrument and its front panel to more fully understand the specifications. It is very easy and informative to mentally tweak the knobs of an instrument even if you can't actually get your hands on one.

Many persons assisted in the preparation of this handbook. Many thanks to Earl Broihier, who helped get the project started and who has provided considerable advice along the way; to Mike Moore and Den Brummell, both of the Heath Company, who spent long hours on numerous picture-taking sessions; and to Shirley Gillham, Gail Bartalone, and Betty Harris, who typed and retyped many versions of the manuscript. A special note of thanks goes to my wife Polly, who not only put up with the whole process but edited each stage of the manuscript.

Charles M. Gilmore

Chapter 1

Electronic Analog Multimeter

Because of the consistently declining cost of the digital multimeter, there is very legitimate concern about the future of the analog multimeter. Despite the analog meter losing some of its popularity to the digital meter, there are still many areas where the analog voltmeter has the edge; the foremost is price. The lowest cost digital meter having 1–1½% accuracy is in the $80 range, whereas $30 still purchases a perfectly suitable electronic analog meter of 2–3% accuracy.

The analog multimeter will be around for some time due to sheer numbers in use. While digital multimeters number in the hundreds of thousands, analog voltmeters number in the millions. That many meters will not be discarded overnight.

The analog meter offers some advantages in operation over its digital counterpart. When the measurement objective is a trend rather than a specific voltage, the analog is the simplest display to interpret. Examples of trend measurements are peaking, dipping, and zeroing (nulling). Many measurements—especially those in the electronic service or maintenance field—are trend, rather than absolute, measurements. The digital meter is especially difficult to use in this mode and can be a source of error and frustration.

Frequently, special scales are required. Special nonlinear scales, such as decibels, are very difficult to achieve with the digital meter but can be provided on an analog meter at little or no added expense. For this reason the person desiring a meter calibrated in decibels is likely to have an analog meter for this purpose.

Many times the analog meter offers a truly portable multifunction meter at lowest cost. Small size is expensive on any meter, but small analog meters tend to be lower in cost than do equally small digital meters. Part of the reason for its inherently smaller size is the normally smaller power requirements of the analog meter.

FUNDAMENTALS

A simplified block diagram of the electronic analog voltmeter is shown in Fig. 1-1. It is essentially similar to its passive cousin, the VOM; however, the electronic analog voltmeter contains an amplifier in addition to the meter and the input attenuator. As noted in Fig. 1-1, the rectifier, which permits AC operation of the voltmeter, may be located either before or after the amplifier.

The amplifier gives the electronic analog voltmeter some distinct advantages. It permits higher input impedance; thus the meter causes considerably less circuit loading. It builds up the input signal so the maximum sensitivity of the voltmeter is not determined by the sensitivity of the meter movement itself. And it permits wider frequency operation than is available with simple meter rectifiers.

Electronic analog multimeters come in three basic types. The original electronic multimeter was the VTVM (vacuum-tube voltmeter). Although the sophisticated electronics man may turn his nose up at the vacuum tube, there are still many VTVMs built and sold today. The VTVM is a moderately accurate, highly stable, high-impedance,

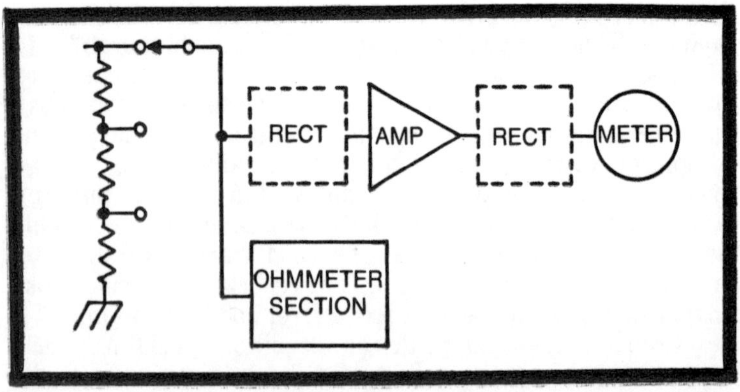

Fig. 1-1. A block diagram of the electronic analog multimeter. The rectifier for AC operation can be placed before the amplifier (VTVMs) or after it (TVMs). The input attenuator here may be either the voltage divider or the current shunts. The ohmmeter circuits supply a test signal to the voltmeter that is proportional to the external unknown resistance.

low-cost voltmeter. To some extent the VTVM has been replaced, especially in the more sophisticated applications, by the TVM (transistorized voltmeter). Most TVMs do all that VTVMs do and more. But there are relatively few truly low-cost TVMs on the market today. The TVOM (transistorized voltohmmeter) is a close relative of the TVM. Frequently the only difference is that the TVM is designed as a bench instrument while the TVOM is designed as a portable instrument. Often the TVOM is referred to as a FET-VOM. This name has developed from the use of the field-effect transistor to obtain the high-input-impedance amplifier. A third class of electronic analog meter is the AC instrument. There are both AC VTVMs and AC TVMs. The AC voltmeter is characterized by the ability to measure AC voltages only, although it can measure very low levels of AC voltage, especially compared to the ordinary VTVM or TVM. Some AC voltmeters also have the ability to display decibels on a linear scale rather than the conventional compressed scale.

VTVM Theory

Figure 1-2 shows two simplified schematics of conventional dual-triode DC VTVMs. Note that the first circuit may be drawn two ways—as a bridge circuit with two of the arms being triodes (A), or as a differential amplifier wth one input referenced to common (B). Alternatively the VTVM bridge circuit may be drawn as shown in C, which, when redrawn in D, shows a pair of cathode followers. Either of these basic configurations may be used to develop the VTVM. The bridge or differential configuration offers a number of advantages. It is relatively free from effects of variations in the power supply, as most variations have equal effect in both arms of the bridge. It provides a simple method of making a zero adjustment, and if the tubes are well matched, it is quite free from effects caused by time and temperature.

The input attenuator of the VTVM is generally a simple voltage divider having a total series resistance of 10M (10 megohms). The accuracy of the divider is dependent on the accuracy of the resistors involved, as the loading from the tube is negligible.

The input impedance of the VTVM is often specified as 11M. This consists of the 10M input inpedance of the attenuator itself plus an added 1M present at the probe tip. The additional 1M, which is shunted by the capacitance of the shielded cable used to conduct the signal from the probe tip to the attenuator, forms a low-pass filter. This low-pass filter keeps AC signals (especially RF) from the input attenuator and out of the tube

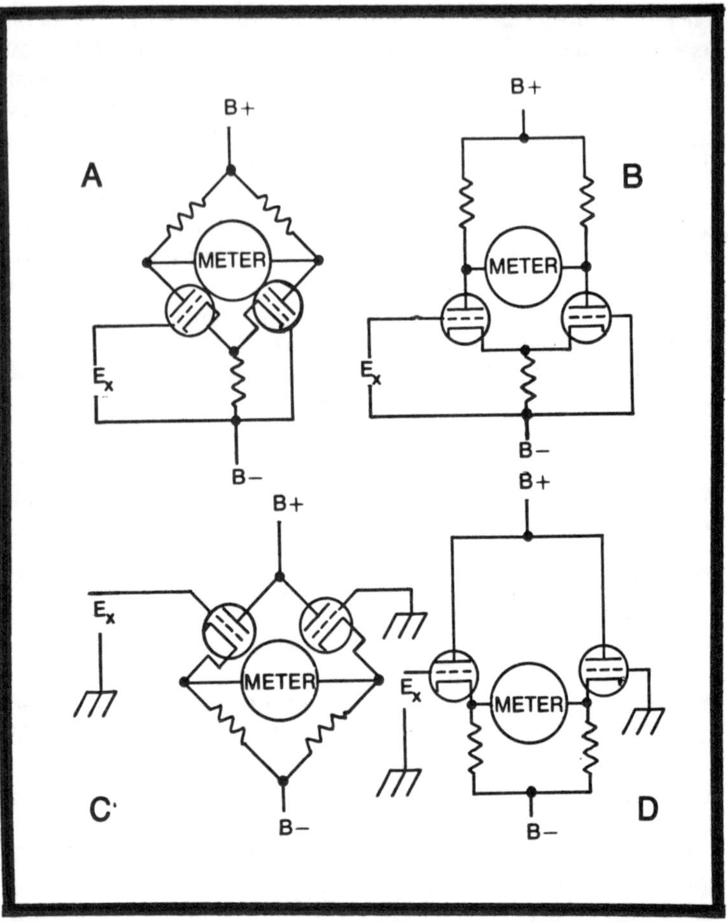

Fig. 1-2. The two basic bridge circuits used in VTVMs. Each bridge is drawn in two configurations for clarity—first as a bridge and then as an amplifier. Parts A and B show the bridge with gain, or the differential amplifier; C and D show the balanced bridge or complementary cathode followers. All configurations offer impedance isolation and long-term stability.

bridge, which is quite sensitive to such signals. When an additional 1M is added to the probe tip, a gain correction must be made to compensate for the voltage divider formed between the 1M resistor and the 10M attenuator.

When the VTVM is switched to AC, a rectifier is inserted in the signal path. This peak-responding voltage doubler is inserted ahead of the bridge, as shown in Fig. 1-3. Most range attenuation takes place at the DC attenuator, after rectification. This technique is used because the greater the

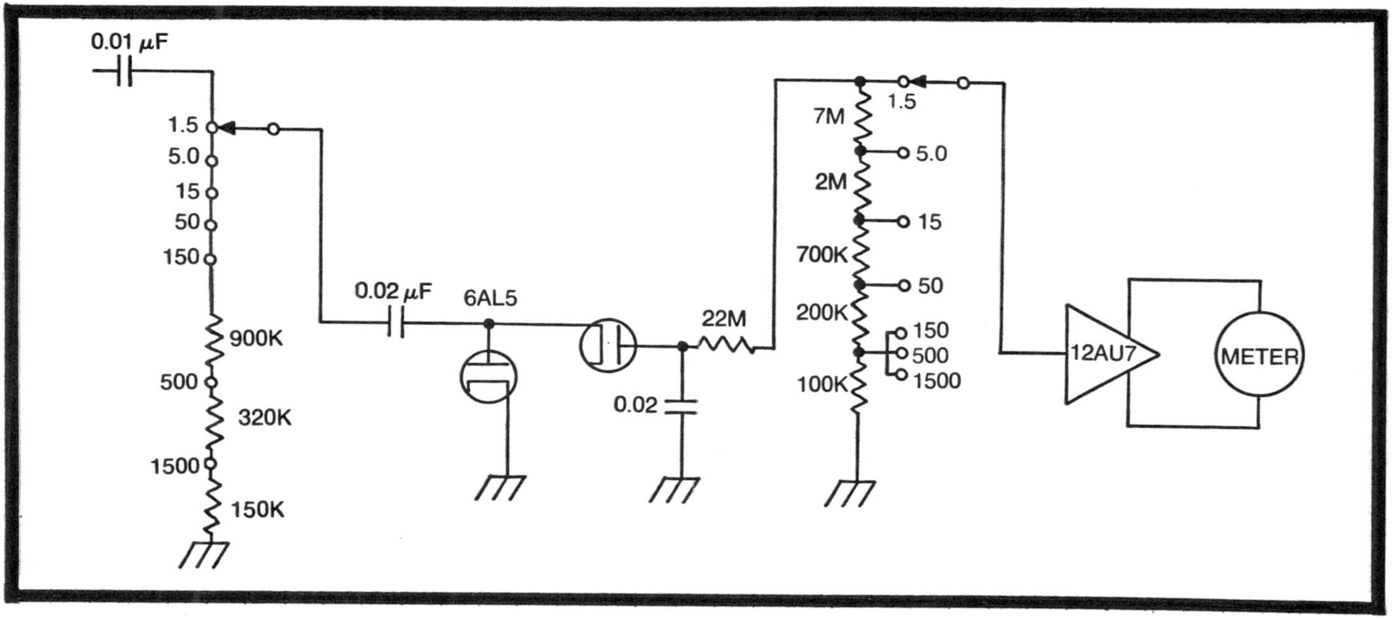

Fig. 1-3. A simplified schematic of the AC portion of the Heath IM-18 VTVM. Note that no attenuation of the AC signal takes place until the meter is switched to 150V. Prior to the 500V range, all attenuation takes place at the 10M DC attenuator following the 6AL5 voltage doubler. The 22M resistor in series with the 10M attenuator corrects the peak-to-peak output of the voltage doubler for RMS indication.

AC signal on the rectifier (without exceeding breakdown voltage), the more linear the conversion to DC. The nonlinearities become sufficiently great at the lowest AC voltage range to require a specially calibrated meter scale. Attenuation takes place for signals in excess of 150V, as higher voltages exceed the peak reverse voltage of the commonly used rectifier tubes. The AC attenuator is not compensated when it is limited to a 1M total series impedance. If either a 10M attenuator or extended high-frequency response is required, the attenuator requires variable capacitors in parallel with the resistors. These capacitors are adjusted to insure that the capacitive portion of the voltage divider is of the same ratio as the resistive portion.

The rectifier of the type shown in Fig. 1-3 has a DC output voltage proportional to the peak-to-peak value of the applied AC voltage. However, most AC measurements are made in terms of the *root mean square* (RMS) value of the signal, which places the work value (RMS) of the AC signal on an equivalent basis with a DC signal. A voltage divider at the output of the rectifier reduces the peak-to-peak DC to a DC voltage equal to the RMS value of the rectified AC signals. Some VTVMs have a mode permitting direct display of peak-to-peak voltage. The RMS value is only accurate if the peak-to-peak/RMS ratio is 2.828, that is, if the input signal is a

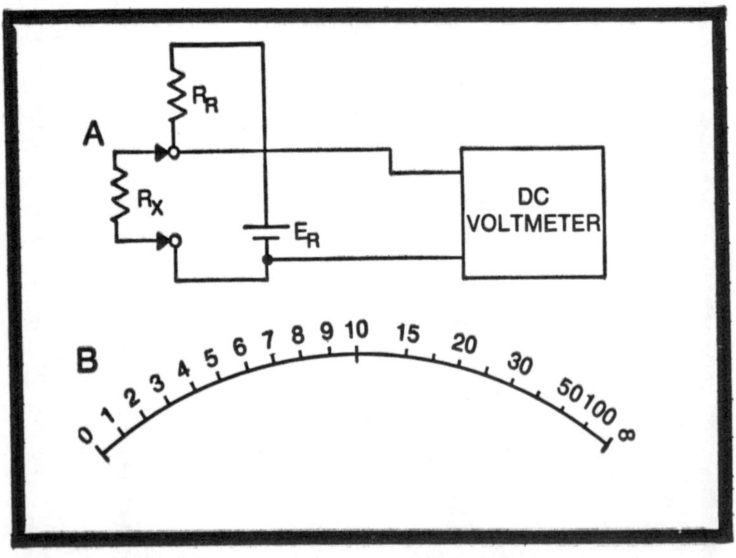

Fig. 1-4. (A) The basic ohmmeter circuit. Voltage E_r is from a 1.5V cell, and on the R_x range, R_r is 10 ohms. The voltmeter scale (B) is calibrated to be proportional to the value of R_x, with 10 ohms at centerscale.

pure sine wave. Other waveforms (square, triangle, and pulse) give erroneous readings on the RMS scale; however, the peak-to-peak scale is correct for these.

The common ohmmeter circuit for the VTVM is the voltage divider type. In this circuit (Fig. 1-4) the meter scale is given a nonlinear calibration. The voltage across the unknown resistance (R_X) is a ratio of the unknown resistance to the sum of the range resistor (R_R) plus the unknown resistor. Typically the lowest resistance range employs a 10-ohm range resistor. With no resistance at the unknown terminals, the voltmeter displays the open-circuit value of the battery. This point is calibrated as infinity on the ohmmeter scale. With a 10-ohm resistor connected to the unknown terminals, one-half the battery voltage appears across the voltmeter. Ten ohms is the centerscale reading of this meter. With a direct short across the unknown terminals, the voltmeter reads zero. When the ohmmeter is ranged to the ×10 scale, the total series resistance is increased to 100 ohms. This gives a centerscale value of 100 ohms. Each succeeding range of the ohmmeter increases the series resistance by a factor of 10. Occasionally other values of series resistance are used.

TVM Theory

The original TVM was nothing more than a solid-state version of the VTVM. With the passage of time, expanded knowledge has changed this. Shown in Fig. 1-5 are three circuits used as the DC portion of solid-state electronic multimeters. Figure 1-5A shows the simple balanced-bridge configuration, which is characteristic of many of the VTVMs. In the solid-state version, the bipolar transistor lacks sufficient input impedance to avoid loading a high-impedance attenuator. Consequently a field-effect transistor is used as an input device. There is no gain present in such a circuit, so TVMs of this type have a maximum sensitivity near 1V.

Figure 1-5B shows the discrete differential amplifier. The advantage of this amplifier lies in its voltage gain. Not only does the amplifier buffer the input attenuator from the low impedance of the meter movement but it also increases the amplitude of the DC signal. Once again the FET is used to obtain the high input impedance. A modern integrated version of the differential amplifier is shown in Fig. 1-5C. An operational amplifier, buffered with FET followers on the inputs, forms the amplifier. The meter is included in the feedback loop of the amplifier. The operational amplifier has the advantage of high loop gain, reducing errors found in the low-gain differential amplifier. Operational amplifiers also

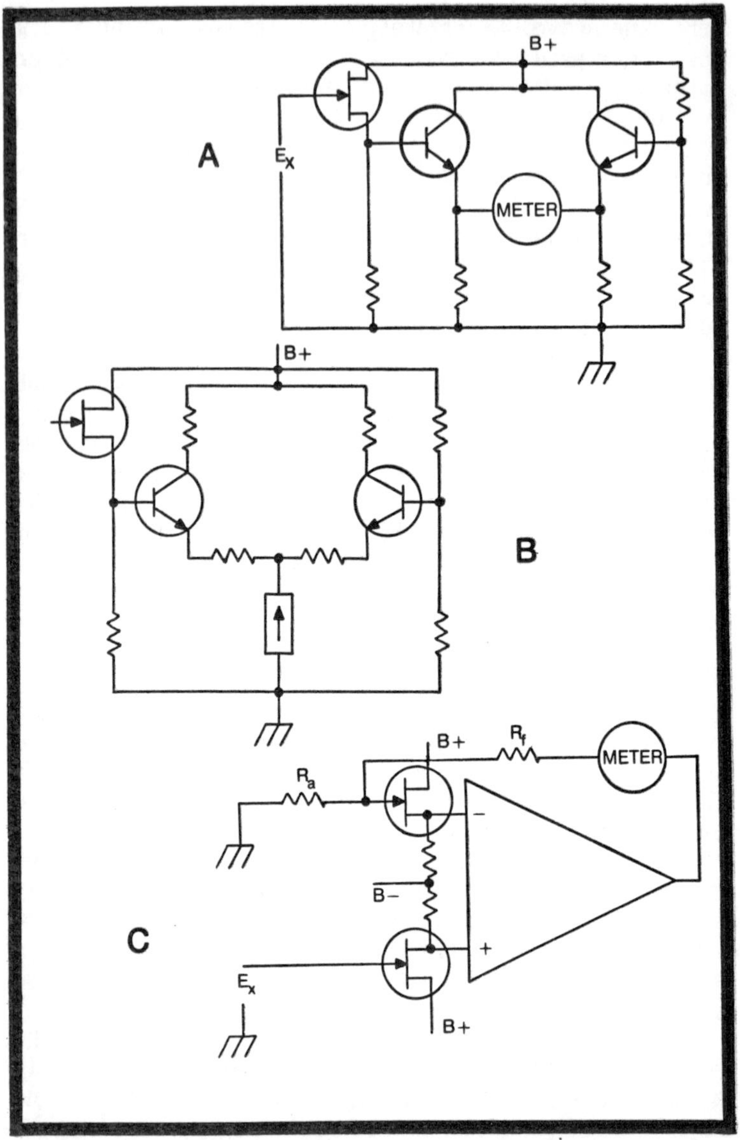

Fig. 1-5. Three different circuits used to form the basic amplifier for the TVM. (A) A simplified schematic of the Heath IM-17 TVM. This circuit is a solid-state version of the classical VTVM bridge of Fig. 1-2D. Note the FET follower, which permits use of a 10M attenuator. (B) A simplified schematic of the Heath IM-25 TVM. This circuit is the solid-state version of the classical VTVM bridge with gain (Fig. 1-2A). Again note the use of an FET to obtain the desired input impedance. (C) A simplified schematic of the Heath IM-104 TVOM. Here an operational amplifier with FET followers in the inputs is used as a voltage follower with gain. Gain is dependent upon the ratio of R_f to R_a.

use low power. which results in a considerable reduction of total parts.

The TVM input attenuator is identical in nature to the simple voltage divider used on the VTVM. Some TVMs use an extra 1M in the probe tip. but the majority of them have a 10M input impedance. And there are a few 1M TVMs.

The AC portion of the TVM takes on a few different characteristics. The placement of the rectifier is the major difference between the VTVM and the TVM. The TVM rectifier is almost always after the amplifier. The amplifier then must have both good AC and DC characteristics. The type of rectifier is also changed. Figure 1-6 shows the various types of

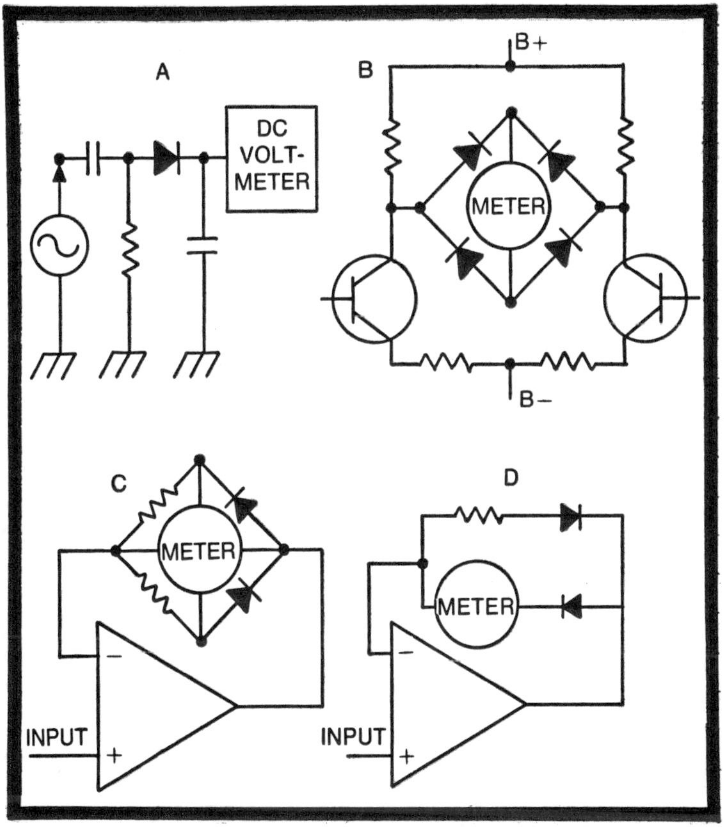

Fig. 1-6. Rectifier configurations used in TVMs. (A) This circuit is used ahead of the amplifier on the very simple TVMs. The rectifier is peak responding. (B) The full-wave bridge rectifier is used after the differential amplifier. The rectifier is average responding. (C) This is one version of the operational meter rectifier. The rectifier is an average-responding, full-wave type. (D) This is a second version of the operational meter rectifier. The rectifier is an average-responding, half-wave type.

rectifiers used with the AC portions of TVMs. Figure 1-6A is the simple input rectifier. much like that on the VTVM; however. this rectifier is not peak-to-peak responding but peak responding. This circuit is used on only the simplest of TVMs. The DC output of this rectifier must be divided by 1.414 to obtain the RMS value of a pure sine wave.

Figure 1-6B is a bridge rectifier driven from the outputs of a differental amplifier. This circuit is average responding. That is. the DC output for a pure sine-wave input is equal to the average value of the sine wave. The output must be multiplied by 1.1 to obtain the RMS value. Figures 1-6C and D are other common implementations of the average-responding circuit. Both of these circuits employ the operational amplifier to overcome the nonlinear characteristics of the diode at low voltages. Figure 1-6C is a full-wave rectifier. and Fig. 1-6D is a half-wave rectifier.

The AC attenuator and the DC attenuator of the TVM are usually the same. in contrast to the normally separate attenuators used by the VTVM. When the attenuator is common. it must be modified to provide accurate attenuation

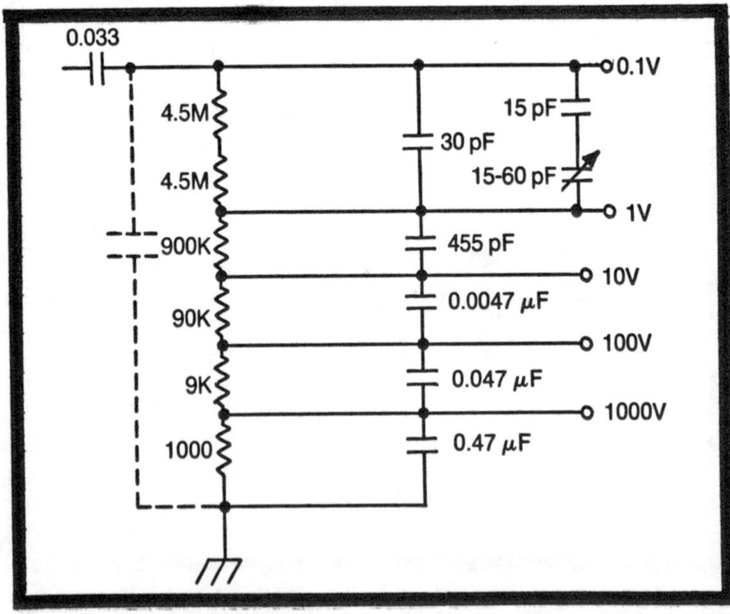

Fig. 1-7. The input attenuator of the Heath IM-104 as arranged for AC operation. Note that this is a decade attenuator, but the instrument has a 1—3—10 range sequence. The additional attenuation is beyond the amplifier. The capacitor shown in dotted lines represents additional stray capacitances from wiring, connectors, etc. The 9M portion of the input attenuator is divided to avoid exceeding the voltage rating of the resistors.

at both low and high frequencies. To do this the attenuator is *compensated.*

Figure 1-7 is a schematic of a compensated attenuator used in the design of a 10M TVM. The capacitors in shunt with the resistances create a capacitive divider with the same ratio as the resistances they shunt. If these capacitors were not used, the capacitive ratio would depend entirely upon the distributed capacitances of the resistors, which are substantial for high resistances and negligible for low resistances. Thus there would be less than a decade attenuation across the first divider and a full decade attenuation across the last divider.

The intent is to have a "47" capacitor across each resistor. On all but the highest value resistors, this is accomplished by shunting the resistor with a capacitor with a "47" basic value. The 900K resistor is only shunted by 455 pF, but an additional 15 pF of distributed capacitance makes up the total shunt capacitance of 470 pF. The 9M resistor (made from two 4.5M resistors in series) is bridged with 30 pF of fixed capacitance plus an adjustable capacitance to make a full shunt capacitance from 37.5 pF to 42 pF. This allows the user (or the manufacturer), when compensating the instrument, to adjust for variations in distributed capacitance from 7 to 9.5 pF. For extreme accuracy the entire divider may have variable compensating capacitors.

In addition to the compensated attenuator the input to the attenuator is also AC-coupled by means of a 0.033 μF capacitor. This capacitor serves to keep DC from the attenuator and therefore from the amplifier and rectifier, which respond to a combination of AC and DC. The total input capacitance of the TVM is somewhat greater than the value of the capacitance in shunt with the divider due to other distributed capacitances (shown by the dotted capacitor in Fig. 1-7.) TVMs using a common AC and DC attenuator have a filter (see Fig. 1-8). This filter is switched into the attenuator output during DC operation and switched out during AC operation. The filter helps remove AC signals from the input to the amplifier when the meter is operated in the DC mode. Although the meter theoretically does not respond to these signals, excessive AC signals can disturb the amplifier sufficiently to cause error in DC readings.

Frequently the coupling capacitor (C_c) defines a low-frequency limit to which the voltmeter may operate. For example, the low-frequency response of the attenuator and input coupling of Fig. 1-6 is 20 Hz for rated accuracy.

The most common form of ammeter on TVMs is the simple shunt type. The TVM offers the advantage of high gain and low

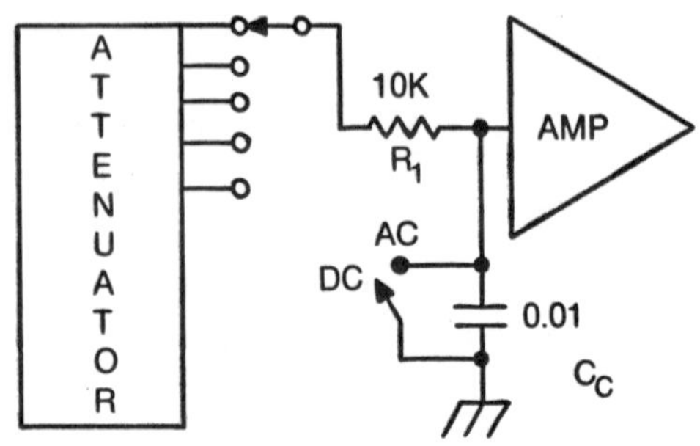

Fig. 1-8. The switchable filter at the attenuator output (amplifier input) when used in the DC mode. This filter has a −3 dB point of 1.6 kHz, thus effectively eliminating RF from the amplifier when making DC measurements.

insertion loss. The VTVM needs a 1V–1.5V insertion loss for a shunt, so ammeters are not normally included. When the shunt type of ammeter is used, the DC and AC ammeters are very similar. The change from AC to DC consists of the selection of the correct voltmeter to be placed across the shunt.

Even though the shunt insertion loss can be kept low with a meter having sensitive ranges (0.1V, for example), the value of the shunt resistance becomes large for the sensitive current ranges. For example, a meter with 0.1V sensitivity has 10K insertion resistance for a 10 μA full-scale range. On the same meter the 1A range has a 0.1-ohm shunt. Due to other series resistances, the 1A range might have an insertion resistance of 0.2 ohms or more.

Ammeters in TVMs less frequently take two other forms. First, consider the scheme shown in Fig. 1-9A. In this circuit an operational amplifier is used as a current-to-voltage converter. The unknown current is introduced to the summing junction of the operational amplifier. This current is nulled by an equal-amplitude but opposite-polarity current developed by the output voltage of the operational amplifier across the feedback resistor. As this current is flowing through the feedback resistor, the voltage drop across the resistor is directly proportional to the resistor and the unknown current. The output voltage of the operational amplifier (which is the voltage across the resistor) is then measured with the voltmeter of the TVM. As there is no voltage drop at the input to the null amplifier, this technique offers the advantage of essentially zero input impedance.

Figure 1-9B shows schematically the technique developed and patented by Weston for current measurement in their model 670 TVM. This TVM features in-circuit current measurements. This is to say, the measurement of current in the circuit does not require opening the circuit, a considerable inconvenience associated with other ammeters. The circuit relies on the inherent resistance in the circuit conductors. The voltage drop across the circuit conductors is reduced to zero by a current equal in amplitude and opposite in polarity to the unknown current. This bucking current is maintained at exactly the correct value by the null amplifier. This technique, which requires a special pair of 2-terminal probes, is

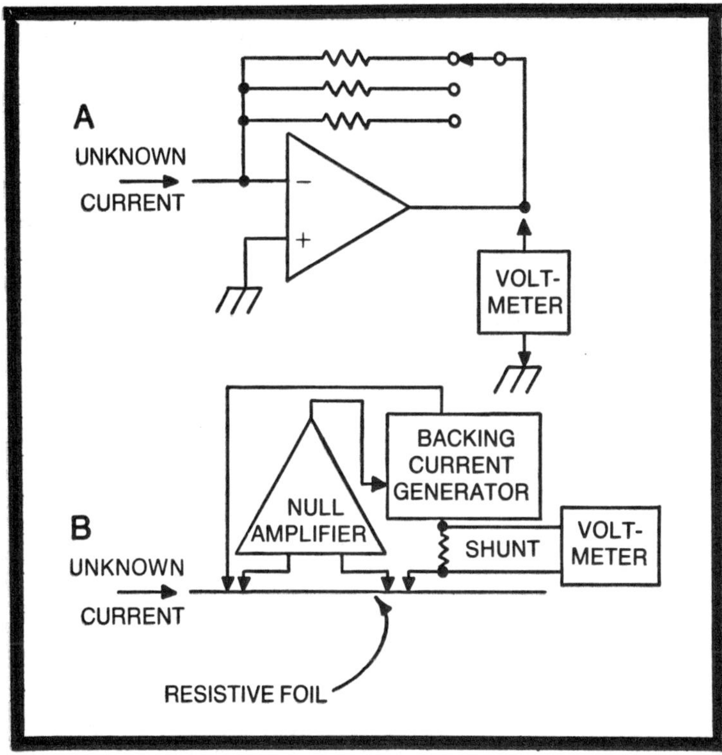

Fig. 1-9. Two unique methods of current measurement. (A) The operational amplifier used as a current-to-voltage converter. Ammeter ranges are selected by the feedback resistor. This method of measurement offers zero input impedance, which is especially advantageous when small currents are to be measured and shunts become high valued. (B) The Weston-patented in-circuit measurement by null detection. The bucking current is made equal and opposite to the unknown current and held at that value by the null amplifier. The bucking current is then measured by conventional means. This ammeter does not require breaking the conductor to make a measurement.

especially useful for measuring currents in PCB foils. Both of these latter techniques are only applicable where the currents being measured are low enough that the measuring circuit can produce an equal and opposite current and where DC currents are being measured.

The simple and most frequently used ohmmeter circuit for TVMs is identical to the series-resistance type of the VTVM. The newer and more "elegant" TVMs feature a dual-range ohmmeter circuit, which has a choice of two open-circuit voltages. A low-voltage range is provided so semiconductor junctions (both germanium and silicon) are not forward-biased. The maximum open-circuit voltage of such ohmmeters is about 0.085V. The high-voltage range of these ohmmeters is usually 1.5V and can easily forward-bias a semiconductor junction.

These two circuits are constructed in two different ways. First and simplest, the low voltage is obtained by reducing the 1.5V ohmmeter battery to the low voltage by means of a voltage divider, which effectively yields a low-voltage source with a 10-ohm series impedance. Alternatively the test signal is derived by the application of a constant-current source to a resistor as shown in Fig. 1-10. In this circuit the

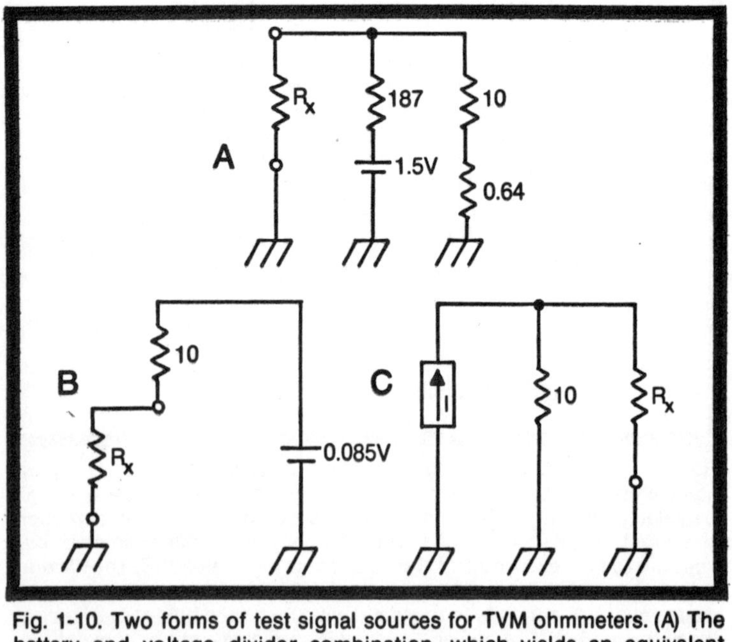

Fig. 1-10. Two forms of test signal sources for TVM ohmmeters. (A) The battery and voltage divider combination, which yields an equivalent circuit (B). (C) The constant-current generator I of either 8.5 mA for a 0.085V test signal or of 150 mA for a 1.5V test signal.

constant-current source is applied to the basic range resistor (say 10 ohms). The value of the current through this resistor determines the maximum open-circuit voltage of the ohmmeter. If the 10-ohm resistance is shunted with an unknown resistance, the voltage drop across the resistances is determined by their parallel value. A 2-range constant-current source is used for high–low measurements. The constant-current technique permits simple control of the maximum open-circuit output voltage of the ohmmeter.

AC Voltmeter Theory

A specialized form of electronic analog voltmeter only amplifies and measures AC voltages. Such a meter is known as an AC VTVM or an AC TVM. The AC voltmeter finds special application in laboratory work and is also found as part of the standard test equipment used in audio service work. The AC voltmeter is characterized by high sensitivity (compared to the VTVM or the TVM), a voltage-only function, extended frequency response, alternate calibration in decibels (usually dBm), and, of course, AC-only response.

The major difference between the AC voltmeter and the other types discussed lies in the amplifier. As DC voltages are not being measured, they need not be amplified. This allows an amplifier design with considerably more gain but without the worries of DC drift which accompanies a high-gain DC amplifier.

The input attenuator of the AC voltmeter is similar to the attenuator described for the AC portion of the TVM. On AC voltmeters with extended frequency response, the input attenuator is often divided into two portions. The first portion is a compensated attenuator providing decade attenuation. Finer attenuation follows an input-buffer stage. This is a lower impedance attenuator and does not need compensation. A specialized form of AC voltmeter offers linear response to logarithmically changing signals. That is, the voltmeter is calibrated linearly in decibels and logarithmically in voltage. Such a meter has an additional logarithmic converter placed between the rectifier and meter circuit. Low-noise amplifiers are required for the AC voltmeter.

SPECIFICATIONS

The voltage ranges of the VTVM are most common in 1.5–5–15 sequence. This sequence, which is a modification of the 1–3–10 sequence, is chosen because the steps represent 10 dB voltage changes. This is convenient when working with AC signals, and most VTVMs have a decibel scale for this purpose. The 0 dB level is 0.774V (1 mW in 600 ohms). Figure 1-11 shows a typical VTVM scale. Occasionally VTVMs are found with a 1–3–10 sequence.

Normally the most sensitive range of the VTVM is 1.5V. The 1.5V AC range may require special calibration to correct for the nonlinearities of the rectifier when operated at this low input level.

The full scale of 150V AC is also handy in countries where a nominal 115V–120V AC power line voltage is used, as measurements of the line voltage are made in the upper third of the meter scale. Of course, this advantage disappears when the common line voltage is 200V–240V AC. VTVMs have an upper voltage limit of 1000V–3000V.

VTVM Input Impedance

Input impedance for the VTVM is generally specified separately for the AC and DC modes. The general-purpose VTVM has an 11M input impedance. Although the 11M input is the most common, there are many other input impedances used. There is little or no significant difference between the VTVM offering 11M input impedance and the VTVM offering 12M input impedance. There is, however, a significant advantage to the VTVM which offers an input impedance in the hundreds of megohms. But such a high input impedance costs lots of money!

If the VTVM is to be used with a high-voltage multiplier probe, then the input impedance of the VTVM with probes disconnected must be some standard value. The accuracy of

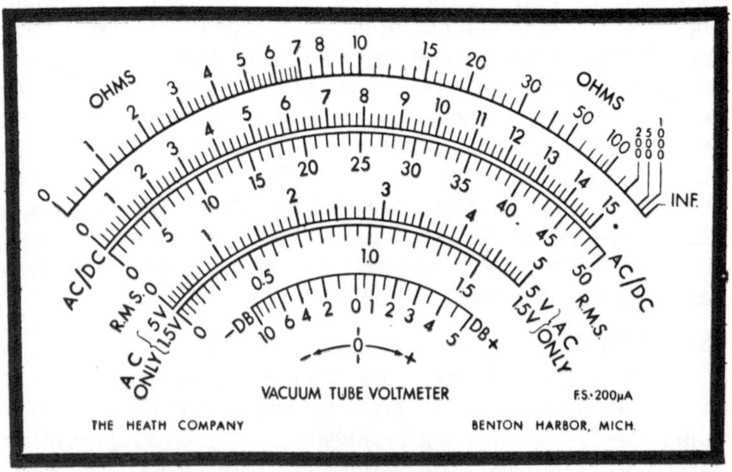

Fig. 1-11. A typical VTVM scale from the Heath IM-28. Note the special ranges other than the simple 15 and 50 scales.

the input impedance is of no consequence unless a multiplier probe is being used. Often the accuracy of input impedance is not specified, or a simple note stating the attenuator uses 1% tolerance resistors is given. Remember, a VTVM which has 11M input impedance with 1M in the probe must be calibrated for that series impedance. The direct input to the VTVM without the probe finds the calibration at these points sensitive by an extra 10%.

The VTVM AC input impedance is 1M, except on the expensive VTVMs, which use the same attenuator for the AC and DC modes. The AC input specifications also include the value of shunt capacitance across the input terminals. This figure does not include the probe capacitance and therefore is simply an indication of the lowest capacitance achievable with special care.

Accuracy of VTVMs

The accuracy specifications of both the AC and the DC modes of the VTVM are given as a percent of full scale. This is a good reason to keep the measurements as close to full scale as possible. Accuracies of ±1% to ±3% are common for the DC ranges, and ±2% to ±5% are common for the AC ranges. The lowest AC range may have a separate error specification because of the nonlinearity of the rectifier at low forward voltages. Accuracy is of course dependent on the price paid for the VTVM . Accuracy diminishes as a result of aging and environmental conditions such as temperature and humidity.

The ohmmeter accuracy specifications are given in one of two ways. On the lower cost VTVMs the accuracy of the ohmmeter is indicated by noting the accuracy of the resistors used in the measurement circuits. This indicates an error about the centerscale value of the ohmmeter. The error at other points must be calculated. The lowest ohmmeter range does not have the indicated accuracy, because of variations in the battery's internal resistance. Accuracy also depends on the preciseness of the ohmmeter scale calibration. The alternate mode of accuracy specification notes the accuracy as a percentage of the midscale reading. VTVMs with such a specification have different specification values for the first range and the other ranges of the ohmmeter. A 2%–5% accuracy is common.

Frequency Response of VTVM

Although the accuracy of the AC mode of the VTVM is given as a percentage of full scale, this accuracy specification must be further qualified by a frequency range. The most common method of specifying frequency response is to indicate the frequency range over which the response will not vary more than ±1 dB when referred to a low-impedance 60 Hz source. The common frequency range for the low-cost VTVM is 25 Hz to 1 MHz. Higher upper-frequency limits can be had for additional cost. The Hewlett-Packard 410B has a 20 Hz to 700 MHz specification, for example, and costs $275.

VTVM Ohmmeter Ranges

As noted earlier, the ohmmeters of the type conventionally used in the VTVM are specified with a centerscale value, usually 10 ohms. The ranging of the ohmmeter is then in decade steps from this point, and ×1M is the highest range. This gives usable measurements in the vicinity of 100M.

Probes For VTVMs

The VTVM is generally supplied with some form of probe. In many cases these probes are either attached to the VTVM or have connectors or characteristics of such a special nature they must only be used with the original model of VTVM. As noted before, the DC probe often has an internal 1M resistance. If one probe is used for all modes of the VTVM, the probe has a switch. Alternatively, separate leads may be supplied for the DC and AC–OHMS functions. There is always a separate lead for the common or return path. Some specialized VTVMs offer an additional probe for extended high-frequency operation. The ease and flexibility of use of the probe are important points to consider when purchasing the VTVM.

VTVM Controls

The VTVM has two front-panel controls in addition to those for selecting range and function. These are the ZERO control and the OHMS adjustment. The ZERO control balances the meter electrically when the test leads are shorted together. This control also can move the meter from the left zero to the center zero, permitting readings of a voltage which passes through 0V. This feature is particularly handy for null detection.

The OHMS adjustment sets the full-scale sensitivity of the voltmeter when used in the ohmmeter function. This feature is necessitated by the variations in the battery used for the ohmmeter voltage source.

The FUNCTION switch provides selection of AC, +DC, –DC, and ohms. The ± selecton of the DC mode is useful, as it eliminates the need to reverse the test leads. Often reversal of the test leads is not practical as it is with the conventional

VOM, as the common test lead may be connected to earth ground. If the common lead is not connected directly to earth ground, the impedance to earth ground (especially the capacitive component) may be quite low. Either situation will cause considerable disturbance to the circuit under test if the voltage source in the circuit is connected to the common lead.

TVM Voltage Ranges

Three different range sequences are common to the TVM. The 1–3–10 presently seems the most common. These meters start at 0.1V, 0.3V, or 1V and extend to 1000V or 3000V. There are a few TVMs which use the 1.5–5–15 sequence, and they cover approximately the same voltage range. When a very-low cost TVM is designed, the scales may run in a 1–10–100 sequence. Such units start with a 1V range and go to 1000V. The 1–3–10 sequence is the most practical when working with voltages found in solid-state circuits.

TVM Accuracies

Accuracy for the TVM, like accuracy for the VTVM, is given as a percentage of full scale and is given separately for the AC and DC voltage modes. DC accuracies are in the ±2% to ±3% area. The accuracy of the DC voltmeter is further limited by environmental factors and possibly the position of the meter. A larger meter aids in taking the reading; and a mirror-backed meter (which reduces reading errors due to viewing angle) is especially helpful for a laboratory, where exacting relative measurements may be needed.

The AC accuracy lies in the area of ±2% to ±5% of full scale, with accuracy being very much related to price. Accuracy is either specified at a particular frequency or over a limited band of frequencies.

Ohmmeter accuracy is much the same as the VTVM ohmmeter accuracy. That is, it has a specification indicating a percentage of the midscale figure. Accuracies are in the area of 5% for TVM ohmmeters. An alternate method indicates the accuracy in degrees of arc of meter swing. This method of ohmmeter specification is one of the most descriptive. An ohmmeter specification of ±3 degrees of arc is common on a meter with 100 degrees of swing. Further limitations are given for the ×1 range and for very low values of resistance.

Frequency Response of TVMs

The frequency response of the AC voltmeter portion of the TVM may be given in a number of ways. The specification may be identical to the AC specification of the VTVM, with AC accuracy specified at a particular frequency (and usually with a low driving impedance) and a frequency response characteristic of plus or minus a number of decibels given for a range of frequencies.

Response limits of ±1 dB or ±2 dB are common for TVM specifications; ±1 dB is approximately ±10% error. The specification may be given as error due to the influence of frequency (e.g., ±5% from 40 Hz to 10 kHz), or the frequency response may be given graphically (decibels vs. frequency).

When limited specifications are given, such as those in a distributor's catalog, the actual numerical values of the frequency response may be omitted. In such cases the meter may only have a frequency range specified or perhaps a simple error specification on the AC ranges. If the latter is the case, this error figure should not be taken as the figure including frequency response, but more as an *idea* of the error at some fixed frequency such as 60 Hz. Note, the frequency response of the TVM is much lower than the frequency response we have come to expect from the VTVM. While the VTVM is specified to 1–2 MHz, the upper limit of all but a few of the TVMs is 10–100 kHz. When buying the TVM, pay close attention to this upper specification of frequency response if you are in need of accurate measurements at the upper end of the audio spectrum. A meter that stops at 10 kHz may not suffice.

TVM Ohmmeter Ranges

The TVM ohmmeter ranges are identical to those of the VTVM. That is, they have a 10-ohm centerscale and ranges from ×1 to ×1M. The TVM has additional specifications for the ohmmeter. First, the maximum open-circuit voltage of the ohmmeter is specified. This is especially true if the TVM has selection of *high* and *low ohms*, in which case a maximum open-circuit voltage is given for both high- and low-ohm operation. The ohmmeter specifications may also give the maximum source current. This specification is useful if there are critical components in the circuits which might be damaged by excessive currents. The common terminal of the TVM is usually negative, but if the manufacturer does not indicate this in the specifications or on the meter itself, it is a good idea to check it out after purchase and note the information where it can easily be seen.

TVM Input Impedance

In the DC mode the input impedance of the TVM is most commonly 10M. There are a few which are still 11M.

The AC input impedance of the TVM is the same DC resistance used on the DC input, with a shunt capacitance specified. The shunt capacitance ranges from a low of 30–40 pF on the very best of TVMs to 100 pF on the more common units. This input capacitance does *not* disappear when the DC mode is used. When measuring the DC component of a signal, this capacitance may cause some degradation of the AC characteristics of the circuit. Unfortunately, few if any manufacturers specify the input capacitance of the DC voltmeter.

TVM Ammeter

Specifications for both the AC and DC ammeter are very much alike. Not all TVMs have the ammeter function, and the lack of ammeter specifications indicates a lack of the function. Ammeter specifications indicate the insertion loss when the ammeter is placed in the circuit. This may be given in two ways. First, the manufacturer may note that the ammeter has a certain maximum voltage drop. Second, the manufacturer may indicate the insertion resistance and give the information necessary to calculate the voltage drop. Some manufacturers give both. The insertion resistance is greater on the high-ampere ranges, as the lead and connector resistances become appreciable.

Accuracy of the ammeter ranges very closely follows the accuracy of the lowest voltage scales of the meter. An additional error figure is often given for the lowest value of shunt resistance (highest current range), as high-accuracy shunts at low resistance are very difficult to make.

Frequency response of the AC ammeter closely follows the frequency response of the lowest range of the AC voltmeter. There may, however, be an additional derating for the high values of shunt resistance (lowest current ranges), as the distributed capacitance and the high value of the shunt resistance cause additional loss of high frequencies.

Ammeter ranges for TVMs run from 10 μA full scale to 1A full scale. A few TVMs offer wider ranges than this; however, this is not common. Ranges up to 10A are offered on some TVOMs designed for appliance servicing.

Probes for TVMs

Normally the probes provided with the TVM are not as elaborate as those supplied with the VTVM. The most common probe is a simple test probe with a test lead terminated in a banana plug or tip plug. Occasionally some TVMs offer a VTVM-type probe. A 11M input impedance is often a clue to a special probe.

AC Rejection of TVMs

A few TVM manufacturers specify the AC rejection of the DC ranges. This specification indicates the number of full-scale overloads of a certain AC frequency (usually 60 Hz) which may be applied without affecting the DC reading.

Protection

As TVMs, and especially TVOMs, are frequently used in situations where accidental contact with high voltages can occur (most commonly either the 120V AC or the 240VAC power line), protection of the DC voltmeter and other functions is important. Protection is accomplished and specified in a number of ways. The manufacturer may simply state the meter is protected from momentary overloads of 220V AC or DC applied to any range. This protection takes the form of back-to-back diodes at appropriate places in the circuit.

Additional instrument protection is given by a set of back-to-back diodes placed across the meter movement itself. These diodes prevent the meter movement from being destroyed by a continuous overload. Ohmmeters are protected by diodes, and a fuse may be used to provide additional protection. Ammeter shunts are often fuse protected. These fuses may be difficult to obtain, and acquiring a few spares is wise.

Some of the TVMs, especially TVOMs, have ruggedized taut-band meter movement. Such an instrument will come through a fall onto a hard surface undamaged or with minor damage, while conventional meters would be seriously damaged.

TVM Scales and Controls

A typical TVM scale is depicted in Fig. 1-12. Note the BATTERY TEST position. The decibel scale is calibrated in decibels referred to 1 mW (0 dBm is 1 mW in 600 ohms, or 0.774V). A small center-zero scale is given for null measurements.

The number of controls and the selection of functions are highly dependent on the functions offered by the particular TVM. Function selection is by rotary switch or pushbuttons. Range selection on the TVM frequency includes a BATTERY TEST position.

A ZERO control is common in TVMs. The range of this control is sufficient to permit zero-centering the meter as well as left-hand zeroing. An ohmmeter adjustment may not be included on a meter which utilizes internal constant-current generators in its ohmmeter function.

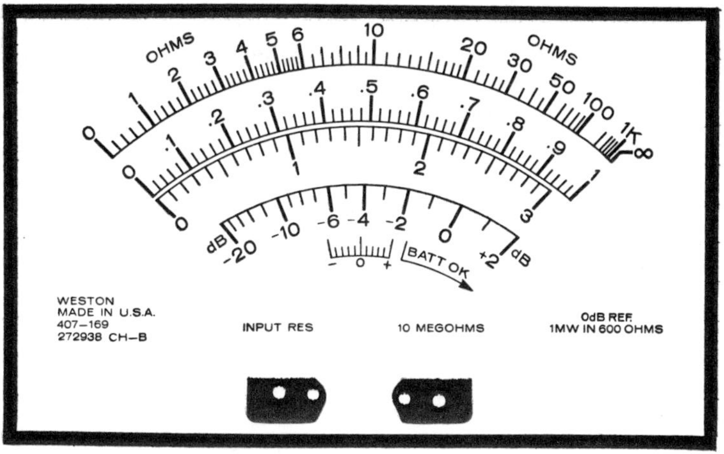

Fig. 1-12. The scales of a Heath IM-104 TVOM. Note the zero-center scale and battery-check position.

Meter Movement

As already noted, some TVMs employ the taut-band meter movement. Others use the moving-coil or D'Arsonval meter. Most manufacturers specify the size of the meter. A 4–5 in. meter is common and adequate for most laboratory and field service applications. Meter displays smaller than this become difficult to read, especially when there are a number of scales clustered on the meter face. Meters larger than 5 in. are useful in some shop or laboratory applications. A mirror-backed meter aids in resolution, but rarely does the TVM offer accuracy requiring such a meter.

TVM FEATURES

Laboratory TVMs offer two different forms of outputs. There may be a DC voltage directly proportional to the meter movement. Such an output may be 100 mV, or 1V for a full-scale meter deflection. This type of output is handy for driving a strip chart recorder or perhaps some form of limit-detecting circuit. The second form of output is a signal taken from the meter amplifier prior to being applied to the rectifier. Such an output may be used to drive a set of headphones, for example, to listen to a circuit as well as to measure it. Such outputs usually have a source impedance of a few kilohms.

Floating Inputs

A floating input is one not connected to the earth side of the power line. Generally this indicates the impedance to ground (usually a very high resistance with a small shunt capacitance) from both the high and low (common) terminals of the voltmeter input is the same. Such a feature allows the voltmeter to be operated without one side of the voltmeter attached to ground. This is particularly handy for measuring voltage drops across circuit elements which have both terminals above ground. A maximum limit is given to this floating capability when the instrument is line-operated. Battery operation inherently provides floating operation.

Accessory Probes

Many manufacturers offer high-voltage probes to extend the useful DC voltage range of the meter and RF probes to extend the useful frequency range of the AC voltmeter. Frequently these probes are not interchangeable from one manufacturer to another. Pay close attention to the specifications of the voltmeter with which they are intended to be used.

Battery Versus Line Operation

As noted above, battery operation can be useful when the instrument is operated with both inputs above ground. Battery operation also eliminates unwanted signals present in a common-ground system. This is desirable when the common input to the TVM is at earth ground. Battery operation is also essential to field operation and is therefore a part of every TVOM. If battery operation is not needed in certain applications, it may be desirable to eliminate them. Batteries do tend to deteriorate after a time and can damage the instrument when they do. Battery operation also adds to the initial price of the instrument.

Special TVM Ammeters

The use of special ammeters is generally dictated by some special application. However, the ammeter in the Weston Model 670, which does not require breaking the foil (of a PCB) to make the current measurement, is especially useful for service work. Ammeter ranges with special high-current ranges are available for special applications. Generally such special scales come at the expense of some other feature or ranges.

Automatic Polarity Indication

Some of the more sophisticated TVMs have polarity indicators instead of a polarity reversal switch. Polarity reversal is indicated by the change of the polarity indicators from plus (+) to minus (−), or vice versa. Polarity indicators are disabled when AC measurements are being made.

TVM Ohmmeter Features

Many TVM ohmmeters offer both *high* and *low* ohmmeter measurements. In addition to the *high* and *low* measurements, some TVMs offer an ohmmeter polarity reversal switch. The switch permits a quick check when forward-biasing a semiconductor junction in a circuit or when testing a semiconductor for shorts or opens.

A few TVMs are available with special ranges. The two most common features of this type are (1) temperature measurement with a special probe and (2) capacitance measurement.

AC Voltmeter Specifications

The AC voltmeter usually has only AC voltage ranges. The voltage ranges cover a wider range than the ranges of either the VTVM or TVM. Ranges extend from the 0.1–10 mV area to the 300V–1000V area. The ranges of the AC voltmeter are also calibrated in decibels.

AC voltmeter *accuracy* is specified in one of two ways. First, the voltmeter may be given an accuracy specification at a single frequency and a frequency-response specification to cover the operating band of frequencies for the instrument. The total operating range of the voltmeter may be subdivided into a number of ranges, with a different response for each range. Second, the voltmeter may be given an accuracy specification for the entire range of operation. Usually this form of specification is reserved for the higher-priced meters. Accuracy may be specified as a percentage of reading plus a percentage of full scale for the AC voltmeter. Again the accuracy may be broken down by frequency ranges.

The *frequency response* of the AC voltmeter is from the 10–20 Hz area to the 1–4 MHz area. The exact extent of response may vary with the voltage range. The frequency response of the AC voltmeter is better than that of either the TVM or VTVM.

An *input impedance* of 10M shunted by 10–40 pF is common for the AC voltmeter. A few AC meters offer a lower input impedance. The input capacitance changes with the voltage ranges selected on some meters. This feature can cause a change in the circuit when the meter is ranged through this point because of a change in capacitive loading.

The *response time* of the AC voltmeter indicates the time for the meter to settle within its rated accuracy. This time is associated with the rectifier circuit. The response time indicates whether the voltmeter may be used as a VU meter, which requires a 0.3 sec time.

Features of AC Voltmeters

AC voltmeters frequently feature amplifier outputs so that the meter may also be used as a high-gain test amplifier. The output impedance of these amplifiers is either 50 or 600 ohms. Other features include an output proportional to the meter deflection (either 100 mV or 1V) filtering that may be switched in to eliminate high-frequency response, battery operation for portability and elimination of ground currents, linear decibel scales, mirror-backed meters for ease in making high-resolution measurements, and a true-RMS converter. Generally such options are only available on higher priced units; and some, such as the true-RMS converter, are quite expensive.

APPLICATIONS

The electronic analog multimeter is used where there is a need to measure a voltage, current, or resistance to an accuracy of 2% or less. This, in fact, is most of the time, especially for the home experimenter or consumer products serviceman. There are industrial servicing and laboratory situations requiring greater accuracy. The digital multimeter should be considered for these applications.

The analog meter is definitely the best display device when trend is the desired information. There are numerous situations in the alignment of an FM receiver or transmitter, for example, where the procedure calls for peaking the output of a stage. An analog meter proves its worth here and should always be carried for this purpose if no other. Many oscillator circuits and some output stages require the supply current to the stage be reduced to a minimum, or dipped. This too is a trend measurement.

The DC output of the FM receiver discriminator is zero when the discriminator transformer is tuned to the center frequency of the IF strip. The zero offset capability of the TVM or VTVM is especially handy for this measurement, as the discriminator output voltage is of one polarity for frequencies above the IF center frequency and of opposite polarity for frequencies below the IF center frequency. Setting the output of the discriminator at zero requires the use of an analog voltmeter. This same nulling procedure is used in a number of bridge circuits, and again, an analog voltmeter with zero offset is called for.

A common problem is the need to make a floating measurement when the meter's common terminal is connected to earth ground. A solution to this is making two measurements, one at each end of the circuit, and subtracting the readings. When this technique is used, the difference between two large numbers is a small number with the same error as large numbers. For example, suppose the voltage drop across R_2 of Fig. 1-13 is needed. Voltage E_1 is measured as 90V and E_2 is measured as 85V. The voltage drop across R_2 is determined to be 5V. Regarding the errors involved in the measurements, the 150V range of a 3% VTVM gives each of the high-voltage measurements a ±4.5V error. This error shifts the floating voltage of R_2 by ±4.5V. The accuracy of the difference between E_1 and E_2 depends on the linearity of the meter. Linearity is not often specified, but 1% of full scale is a good figure to use. The 5V drop across R_2 as measured by this technique can therefore vary by as much as ±1.5V ($150V \times 0.01 = 1.5V$).

Voltmeter Selection

Many alignment and test procedures call for the use of a specific type of voltmeter, frequently the VTVM. The 10M TVM is a good substitute if the circuit is known not to need the isolation of the 1M probe resistor of the VTVM. It is wise not to use a DVM for such measurements. Often the added resolution of the digital meter causes needless confusion. If a 10M TVM is the only instrument available, a special probe can be made. This probe with the additional 1M resistance at its tip reduces

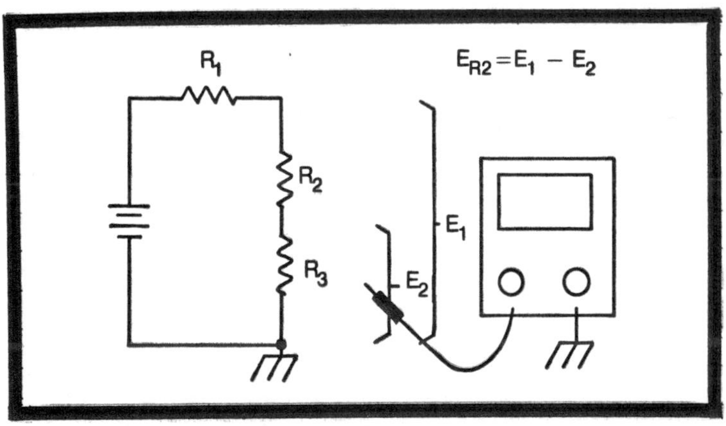

Fig. 1-13. The voltage across R_2 cannot be measured with this electronic voltmeter, as the common lead is tied to earth ground along with the circuit common point. The voltage E_{R2} must be determined by two measurements and a calculation.

all measurements by 9%. However, the measurements are still valid, since the error is *known*. Figure 1-14 depicts the construction of such a probe.

The in-circuit current meter is useful for troubleshooting. For example, assume one of 10 integrated circuits soldered to a PCB is shorted internally enough to make the power supply enter current foldback. The normal method of servicing is to replace or disconnect each integrated circuit from the power rail until the power supply recovers. With the in-circuit current measurement, the current to each integrated circuit is measured. The shorted device is identified as the one drawing a large current. Only the defective component is replaced, saving considerable work. This technique is applicable to many similar situations in both discrete and IC designs.

Errors

The analog voltmeter, like any other instrument, is capable of introducing errors in the circuit being analyzed. These errors are fundamentally of two types—those introduced by the operator and those caused by the instrument.

Errors in reading are the most common caused by the operator. Reading the wrong scale or range is quite common, especially with the crowded scales of a multifunction meter. For example, it is easy to have the voltmeter set on the 15V range and take the reading from the 50V scale. A more subtle mistake is to take a 1.5V AC reading from the 15V scale rather than the special scale provided for the 1.5V AC range.

Errors in reading can also be generated by *parallax*. Parallax is the difference in readings obtained by viewing the meter from different angles. Generally, errors due to parallax are only present when high-precision measurements are being

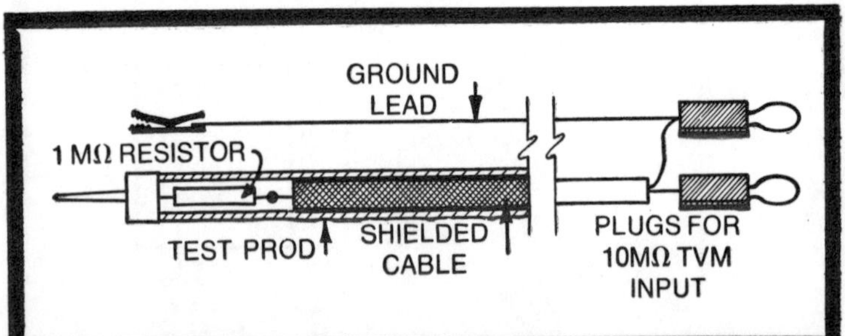

Fig. 1-14. An isolating probe for 10M TVMs. This probe effectively places a low-capacitance 1M load on the circuit being measured instead of the 10M high-capacitance load offered by the TVM. Warning: This probe will make all DC voltage readings low by 9% unless the TVM is recalibrated for the probe (not recommended).

made. Where parallax is a severe problem, a mirror-backed meter scale helps maintain all readings from a single point.

Another operator error is to make measurements with the probe tip switch in the wrong position.

Resistance measurements made with current still flowing in the circuit are invalid. The meter, in the final analysis, measures voltage, and any extra current in the resistance being measured causes extra voltage, which is a source of error. If the meter is not protected, there may also be damage to the ohmmeter circuits.

Resistance measurements made on transformer windings (or other highly inductive components) are susceptible to two problems. The first is caused by ignorance. The operator discards a good transformer because the resistance of a winding appears to be low. What has been forgotten is the *reflected* impedance from the other windings. The second problem, caused by the high inductance itself, is normally only a concern with power transformer primaries and high-voltage secondaries: When the ohmmeter is disconnected from the winding, the current loss causes a collapse in the magnetic field. This collapse causes a high voltage to appear at the winding terminals, which can cause electrical shock, damage to the ohmmeter, or both.

Polarity errors are common and come in two forms. First, the user may have the meter set for one polarity and assume another. The second polarity error is associated with ohmmeter measurements. In this situation the operator fails to consider the effects ohmmeter polarity may have on a circuit containing semiconductor junctions. If the presence of semiconductor junctions in the measurement is suspected, a simple check is the reversal of the test leads. Significantly different resistance readings indicate the presence of such junctions.

Improper interpretation of the decibel scales often leads to error. (Refer to the appendix for a comprehensive table of decibels vs. current, voltage, and power ratios.)

As to errors introduced by the meter itself, circuit loading is probably the most common. Often a simplistic view of the VTVM or TVM is taken, with the assumption that the electronic voltmeter does not load the circuit at all. Although the electronic voltmeter may be substantially better than the VOM in this regard, it still loads the circuit. In the simple DC case a 10M voltmeter loads a 1M source sufficiently to produce a 10% error in the reading. This is if DC only is taken into consideration.

If the AC component of the signal is considered, capacitive loading must also be considered. The meter input capacitance often presents a low-impedance load compared to the high-impedance resistive load of the divider. As noted earlier, this must be taken into consideration when making DC measurements as well as AC measurements, especially when dealing with a circuit where the loss of AC shifts the DC values of the circuit. The self-biasing stages of a transmitter are an example of such a case. The isolating resistance of the VTVM is necessary in such cases.

There are times when the VOM is actually better than the TVOM or the VTVM. For example, the VTVM has an 11M input impedance when used on the 1500V range, but a VOM with a 20 KV input impedance has a 30M input impedance ($20 \times 10^3 \times 1.5 \times 10^3 = 30 \times 10^6$). When measuring a high-voltage supply, this may be of considerable advantage. The VOM may have the further advantage of having an even higher voltage range (4–5 kV for example), and the circuit loading caused by these ranges is even less than that of the VTVM or TVM.

Failure to account for the effects of a shunt when making current measurements may lead to measurement error. This becomes even more true as measurements are made of low-level currents. When low-level currents are being measured, shunts of many kilohms are used. The introduction of any additional 1K–10K to the circuit may have a serious effect upon its operation.

SUMMARY

The analog voltmeter is truly not dead. Although many of its functions are being replaced by the digital multimeter, there are still many places where the analog instrument is entirely adequate and is definitely the more economical instrument. The electronic analog multimeter is continuously improving. Meters available today are very sophisticated and make complex and accurate measurements. However, not to be forgotten is the VTVM, the last vestige of vacuum-tube technology in electronic instrumentation and still the best value available in instrumentation.

Chapter 2
Voltmeter

Rarely do electronic instrument users have the good fortune that an instrument becomes so much in demand that mass production techniques can be used in its manufacture, lowering the price enough that it becomes a true bargain. The DMM (digital multimeter) is one of those instruments. High user demand has brought sophisticated technologies to produce a field instrument for a few hundred dollars which just a few short years ago was confined to the temperature-controlled laboratory and cost thousands of dollars.

The concept of the digital voltmeter[1] has been around for some time. Nonlinear Systems introduced the first digital voltmeter in the early 1950s. This unit would only measure DC voltages, used a combination of vacuum tubes and relays, was not exactly portable, and cost thousands of dollars. But it was a start. From that point to the mid 1960s the instrument remained *the* laboratory digital voltmeter. Major changes were made in the accuracy, type of conversion (analog to digital), and resolution; and circuit components changed from vacuum tubes, to discrete solid-state, and then to integrated circuits.

[1] The terms *digital voltmeter* (DVM) and *digital multimeter* (DMM) are used interchangeably. Strictly speaking, a voltmeter measures only voltage, but DVMs as well as VTVMs measure other quantities also. In this discussion the word *voltmeter* is to be taken literally, and a DVM is to be considered a voltage-only instrument.

The DMM entered with the advent of commercial integrated circuits and the profusion of electronic equipment in the field. The first units barely classified as portable and their cost was nearly a thousand dollars. But the advances were rapid. In the early 1970s the DMM as we know it was brought to the market. Ever since, there has been a great race to see who could build the lowest cost, most accurate, highest volume DMM. The result is hundreds of thousands of these instruments for many applications—for the user, really a dream come true.

FUNDAMENTALS

The basis of the digital multimeter is the technique by which the instrument converts analog signals to digital ones. Once an analog-to-digital (A–D) converter is available, the multimeter can be created in much the same manner as the analog VOM can be made with a DC meter movement.

The vast majority of DMMs use one type of analog-to-digital conversion: *dual slope*. There are, however, two other types which have been used to construct digital voltmeters. The *single-slope* converter is used to generate a very low cost instrument. It has the advantage of simplicity, so the theory of the single-slope converter will be explained first. The *voltage-to frequency* (V–F) converter will be discussed last. Although not so common today, it does have some special characteristics which make it interesting, and there are numerous instruments in the field employing this conversion technique.

Single-Slope Converter Operation

The fundamental concept of the single-slope converter is simple. It is to convert an unknown voltage to time. Time is chosen because simple digital counting circuits and an oscillator of known frequency may be used to display time in a digital format. To make the single-slope converter work, it is necessary not only to convert the unknown voltage into time but to do it in a linear manner.

To convert voltage to time is relatively easy. If a voltage is used to charge a capacitor through a resistor, the voltage across the capacitor will increase with time and approach the charging voltage. The time required for the voltage across the capacitor to reach a particular value will depend on the value of the voltage source, the capacitance, and the resistance. Unfortunately this charging rate is not linear but exponential.

The charging of a capacitor may be made linear, however, if the source of the charging current is a constant current instead of a constant voltage with a series resistance. Figure 2-1 shows three curves of voltage across a capacitor versus time. Section A shows the familiar exponential charge obtained from a constant-voltage source with a series resistance. Section B shows the linear charge obtained from a constant-current source. Section C shows a capacitor being discharged from some initial voltage by a constant-current source (source of negative current as the current flows out).

Figure 2-2 is a block diagram of a simple single-slope converter. The voltage across the capacitor may be held at 0V by the shorting switch S_1. The voltage E_C across the capacitor, generated by charging the capacitor from the constant-current source, and the unknown voltage E_X are applied to separate inputs of a voltage comparator. The voltage comparator converts analog information at its input to a digital signal at its output. If the voltage at the + input of the comparator is greater than the voltage at the − input, the output will be a logic 1. However, if the voltage on the + input is less than the voltage on the − input, the output will be a logic 0. The output of the comparator is connected to one input of a gate. The other input of this gate is connected to an oscillator of known frequency. When the output of the comparator is a logic 1 and the switch S_2 is at a logic 1, the oscillator signal passes from the gate to the counting circuits; however, if the output of the comparator is logic 0 or switch S_2 is logic 0, no signal is passed to the counters.

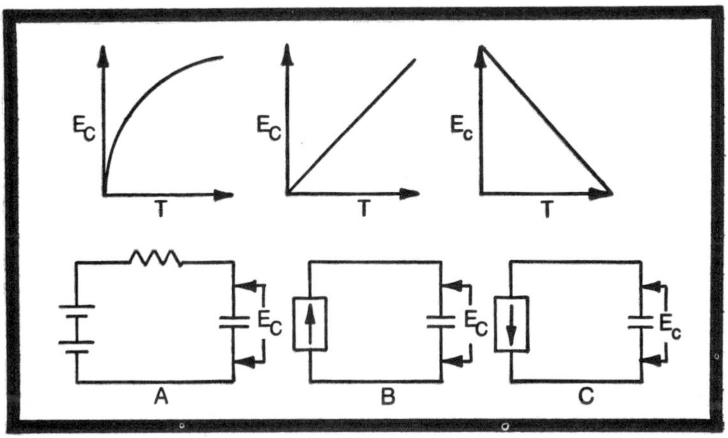

Fig. 2-1. Three curves showing the voltage across a capacitor versus time. (A) The exponential voltage across a capacitor being charged from a constant-voltage source (a battery) through a series resistance. (B) The linearly increasing voltage across a capacitor being charged from a constant-current source. (C) The linearly decreasing voltage across a capacitor being discharged by a constant-current source.

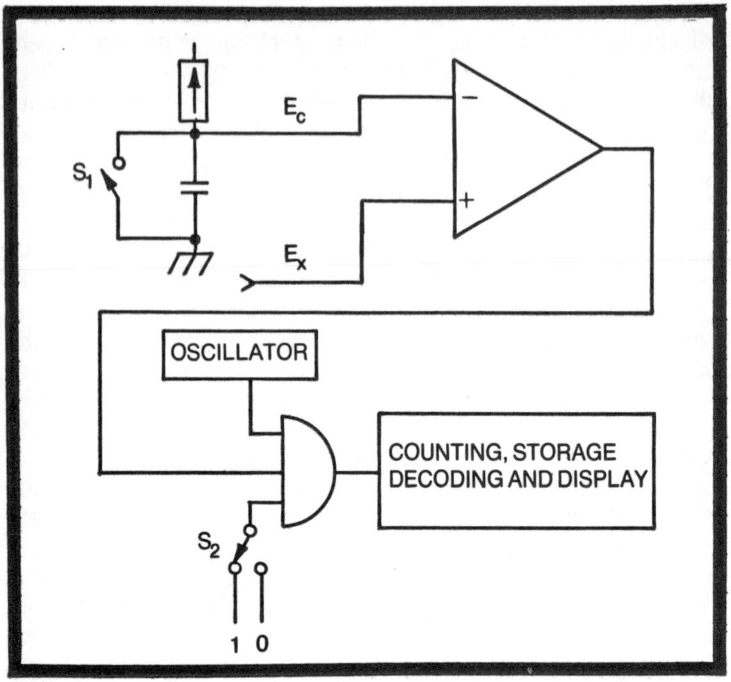

Fig. 2-2. The single-slope analog-to-digital converter. The linearly increasing voltage on the charged capacitor E_c is compared to the unknown voltage E_x by the comparator. The comparator permits signals from the oscillator to be counted during the time required for E_c to charge to a value equal to E_x. The count is directly proportional to voltage E_x and is displayed on the 2- or 4-digit display.

Initial conditions of operation require the counters to be reset to zero, S_1 closed, S_2 at logic 0 (keeping the charge on the capacitor at zero and blocking any oscillator signal from passing through the gate), and that an unknown voltage be placed on the − input of the comparator.

The measurement cycle is initiated by simultaneous opening of S_1 and switching S_2 to logic 1. When S_2 is switched to logic 1, both control inputs of the gate become logic 1, as the unknown voltage at the + input of the comparator is greater than the initial voltage across the capacitor. With both of the control inputs at logic 1, the oscillator signal passes to the counters, which begin to accumulate counts. Simultaneously the voltage across the capacitor begins to rise, as there is no longer a short across it diverting the charging current. After some length of time has passed, the linearly rising voltage across the capacitor becomes equal to the unknown voltage at the + input of the comparator. An instant after these voltages become equal, the voltage at the + input is less than the voltage at the − input. The output voltage of the comparator therefore switches from a logic 1 to a logic 0. This change on the output of the comparator causes the oscillator signal to stop flowing through the gate to the counters. The closing of the gate terminates the measurement.

The counts now accumulated in the counter represent the time the gate was open. The time the gate was open is directly and linearly proportional to the unknown voltage. Therefore the counts in the counter represent the unknown voltage. The final step is to modify the charge rate of the constant-current source or adjust the oscillator frequency so the accumulated value in the counter is proportional to the unknown voltage and so 1.000V produces exactly 1000 counts.

There are, of course, limits to such a simple converter. First, it only measures voltages of a single polarity. Second, circuitry must be added to detect input voltages which are greater than the maximum charging voltage of the capacitor. (The latter is known as an *overrange* condition.)

Long-term errors tend to creep into the single-slope converter. First, the converter is susceptible to drifts in the oscillator frequency. Second, it is susceptible to drifts in the value of the output current of the constant-current source. Third, accuracy depends on the stability of the capacitor. Fourth, accuracy depends on the stability of the differential voltage required to trip the comparator. Fifth, the converter is highly susceptible to any noise which might be present on the unknown voltage, especially as the voltage on the capacitor approaches the unknown voltage. As in any charge-type converter, the accuracy of the converter depends on the linearity of the capacitor's charge rate. The linearity requirement necessitates a good capacitor and a PCB layout free from high-resistance leakage paths. Glass epoxy circuit boards are usually chosen for this reason.

Even with the aforementioned and other faults, the single-slope converter has been used to produce a number of satisfactory low-cost DMMs. The simplicity of the circuit makes it very appealing.

Dual-Slope Converter Operation

The need for improved accuracy, especially long-term accuracy, forced the development of an improved converter. This converter uses the linear-charging concept of the single-slope converter but adds improvements to make the long-term stability independent of the many variables plaguing the single-slope design. This new design is called the *dual-slope* converter.

The dual-slope converter also utilizes a capacitor charged by a constant-current source to provide a voltage—time conversion. The dual-slope converter differs from the single-slope type in that extensive effort has been made to reduce the accuracy and stability requirements placed on components and subsections of the converter. The result is a converter that maintains high accuracy (usually in the 0.5% to 0.01% region) over extended periods of time, with wide variations in temperature, and with high input noise. The design also uses low-cost components.

There are many variations on the dual-slope conversion technique, but a basic technique follows. The dual-slope converter, as its name implies, does not charge or integrate in only one direction, but charges the capacitor and then discharges it during a measurement cycle. This charge—discharge cycle of the dual-slope converter tends to eliminate one of the major problems of the single-slope converter—i.e., long-term inaccuracies brought about by the slow drift of the components involved in the charging circuit.

Referring to Fig. 2-3, a block diagram of an elementary dual-slope converter, these major subsections can be identified:

1. *An operational amplifier connected in the integrating configuration.* When a negative voltage is applied to the resistor *R*, the voltage at the output of the operational amplifier and thus across the capacitor will rise linearly in time. In effect, the capacitor is being charged by a constant-current source.
2. *A switch* S_1 *which can connect the integrator input to either the unknown voltage* V_X *or to (3).*
3. *A reference voltage* V_{REF}. This switch is controlled by the converter circuitry.
4. *A comparator connected to the output of the integrator, which has a reference voltage (usually 0V) connected to its other input.*
5. *A gate connected to the comparator output.*
6. *An oscillator and counter circuit identical to those in the single-slope converter.*
7. *An overrange circuit connected to operate the input selector switch* S_1.

The dual-slope converter's initial conditions at starting time must be: counters reset to all zeros and input switch in the E_X position. Initially the output voltage of the integrator

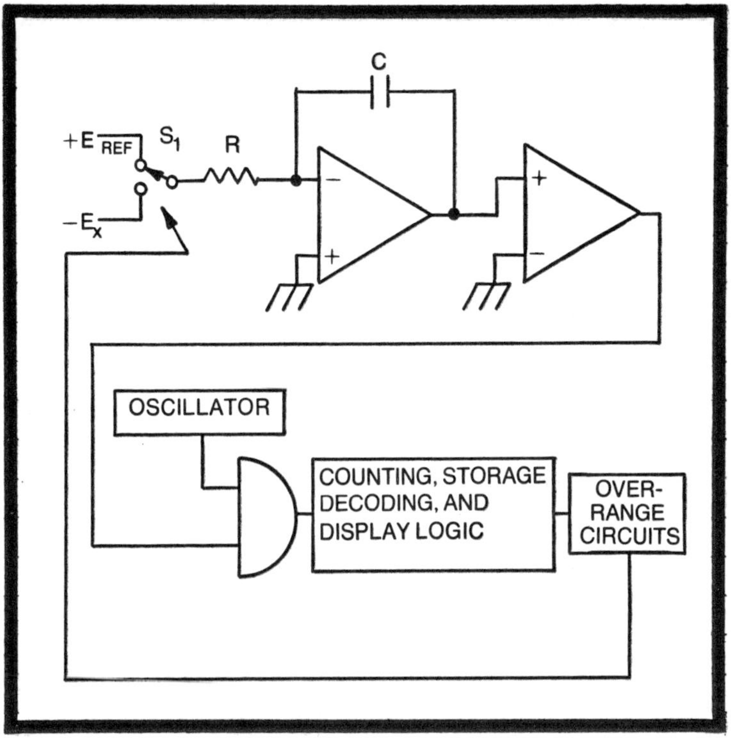

Fig. 2-3. A simplified block diagram of a dual-slope A–D converter. Switch S_1 is a high-speed analog switch actuated by the first overrange of the counting circuits. The first overrange is not displayed.

will be at a somewhat negative value from the previous conversion. (See Fig. 2-4 for timing diagrams.) This negative output causes the comparator to have a logic 0 output, keeping the oscillator signal from the counter circuits. As soon as an unknown negative voltage E_X is applied to the converter input, the integrator begins to charge. Voltage at the integrator output increases in a positive direction during the time t_0 to t_1. At t_1, the output voltage of the integrator passes through 0V. The comparator detects this change; the comparator output stage changes to a logic one, thus permitting the counter circuits to begin accumulation of counts from the oscillator. The integrator continues to increase its output voltage until the switch is moved from the E_X position to the E_{REF} position at t_2. The switch is actuated by the first overrange of the counters. In other words, the time the integrator is allowed to charge after the output passes through 0V (t_1 to t_2) is the time required to completely fill up the counters once.

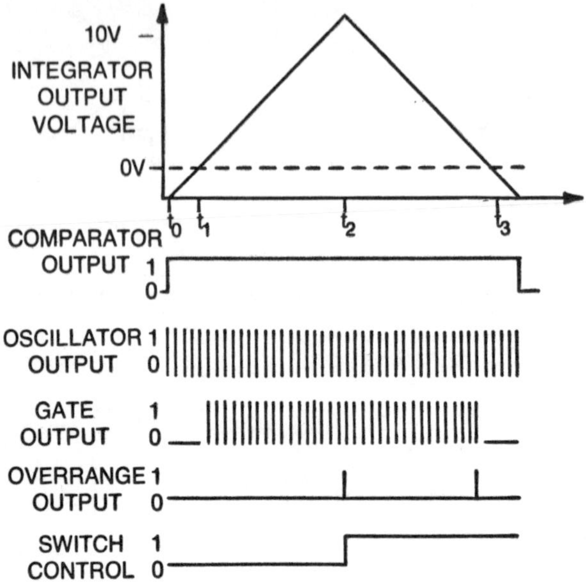

Fig. 2-4. Timing diagrams for the dual-slope converter operated with the input voltage E_X equal to the reference voltage E_{REF}. Note that the charge and discharge times are equal (e.g., t_1 to t_2 equals t_2 to t_3). Thus the counts accumulated in the discharge cycle are identical to those which established the time of the charge cycle.

The output voltage of the integrator at the time the counters go into overrange is determined by a number of factors: (1) the values of R and C, which are fixed, or at least presumed to be constant for the entire conversion cycle; (2) the time allowed for integration, or in other words, the time required for the oscillator to fill up the counter circuits and overrange them (the frequency of the oscillator is also presumed to be stable for at least the duration of the conversion cycle); and (3) the value of the unknown voltage.

Once the switch has been connected to the reference voltage, the output voltage of the integrator begins to decrease. The rate of decrease is determined by the value of resistance R and capacitance C of the integrator and by V_{REF}. The reference voltage is a very closely held value in the dual-slope converter; both its short-term and long-term stabilities must be good.

After a sufficient time has passed, the output voltage of the integrator passes through 0V at t_3 and continues to a negative value. When the output voltage of the integrator passes through 0V at t_3, the comparator detects this condition and disconnects the oscillator signal from the counting circuits. The status of the digital counters at this time reveals a count directly proportional to the ratio of the input voltage to the reference voltage. For example, if the reference voltage is exactly equal to the unknown voltage, the counter just overranges for a second time at the instant the comparator shuts off the oscillator signal. If, however, the unknown voltage is less than the reference voltage, a count less than the full value of the counters remains.

Suppose the reference voltage is 1V, the oscillator frequency is 1 kHz, and the counter has three decades. For the first situation, assume a 1V signal at the unknown terminal (see Fig. 2-4). Between times t_1 and t_2, 1 sec passes, as 1000 pulses from the oscillator are required to overrange the 3-decade counter. At overrange assume the integrator voltage to have charged to 10V. If 1 sec was required to charge the

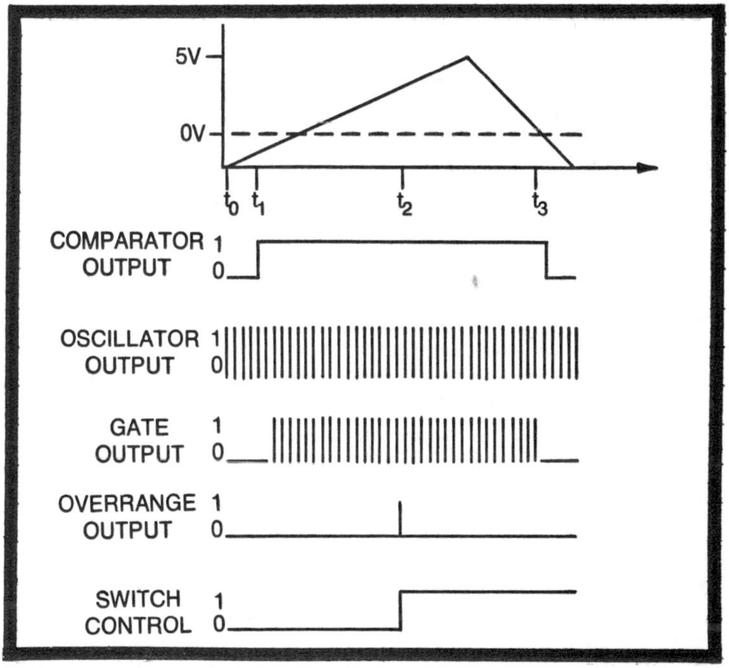

Fig. 2-5. Timing diagrams for a dual-slope converter operated with the input voltage E_X one-half the value of the reference voltage E_{REF}. Note that the larger reference voltage causes a faster discharge rate than the charge rates caused by the smaller input voltage. This condition allows the counters to fill to only one-half of the number of counts which established the time of the charge cycle.

capacitor to 10 V by a 1V signal, then 1 sec additional will be required to discharge the capacitor to 0V with another 1V signal. The counters will start (time t_1) at 000; they will be 1000 at t_2 and 2000 at t_3. However, the first overrange will not be displayed, and the 2000 count will be displayed as 1000, or 1V.

If the unknown signal were 0.5V, the following changes to the above example would have to be made (see Fig. 2-5). The output voltage of the integrator at the end of the first second would only be 5V, not 10V, as the source would be only one-half as great. Discharge would be at the rate of 10V per second; but with only 5V to be discharged, only 500 msec of discharge time would be required. The sequence of the counters would be: 000 at t_1, 1000 at t_2, and 1500 at t_3. As the first overrange is not displayed, the displayed count would be 500, or 0.500V.

If the charge and discharge rates of the integrator are linear (and this is a fundamental requirement), the relationship described above will hold true for any ratio of unknown to reference voltage.

The dual-slope converter improves upon some of the major problems in the single-slope converter by eliminating dependence on many of the tightly controlled parameters of the single-slope converter.

In the single-slope converter, accuracy of the conversion depends on the long-term accuracy of the voltage-to-time conversion of the integrator and thus on the stability of *R* and *C*. In the dual-slope converter the requirement is only that the *RC* product of the integrator be constant for any one conversion cycle. That is, the *up-ramp RC* product and the *down-ramp RC* product must be identical. However, there is no requirement that the *RC* product of one conversion cycle and that of another be identical.

In the single-slope converter, long-term stability of the oscillator is a requirement. In the dual-slope converter the oscillator must be stable only for the length of one conversion cycle. Again, stability from conversion to conversion is not of importance to the fundamental accuracy of the converter. The dual-slope converter only depends on the ratio of the *up* time to the *down* time. Therefore an oscillator 5% high in frequency simply shortens both the *up* ramp and the *down* ramp by 5%, leaving no net error.

The single-slope converter requires stability of the comparator trip point. If the differential voltage between the two comparator inputs should drift (commonly this changes with temperature), the voltage measured by the single-slope converter changes. With the dual-slope converter, drift in the comparator trip point does not cause a problem unless it happens between the time of the *up* ramp and the *down* ramp of a conversion cycle.

Noise Rejection of Dual-Slope Converter. The dual-slope converter also has some inherent noise-rejection capabilities. The most common noise problem, apparent when making DC measurements, is caused by line-frequency signals riding on the DC voltage to be measured. Considering the single-slope converter, it is possible that noise on the input line will cause the comparator to trip on the sum or difference of the noise and the DC input signal. For example, a 0.5V DC signal having 0.25V of AC peak noise could cause a measurement as high as 0.75V or one as low as 0.25V.

The input to a dual-slope converter is applied to an integrator. The output voltage of the integrator at the end of the *up* cycle is the average value of the DC signal, plus any noise which might have been present during the integration period. The length of the integration cycle is determined by the number of counts required to fill and overrange the counter and by the frequency of the clock.

The length of this integration is of course constant from conversion to conversion within the stability of the clock. The average value of a sine wave is zero after both the positive and negative halves of a cycle are complete. The oscillator frequency of most dual-slope converters is adjusted to make the unknown integration cycle a whole number of line-frequency cycles. Typically the integration time is chosen to be an even multiple of the period of one cycle of the line frequency—for example, 100 msec, which is the time required for six cycles at 60 Hz or five cycles at 50 Hz.

Integrating six cycles of the interfering signal improves the technique of averaging out sinusodial noise. This technique gives the dual-slope converter a good immunity to line-frequency-related noise. The rejection of such signals is called *line-frequency normal-mode rejection.*

Overranging. Obviously this simple converter needs some supporting circuitry to create a full-fledged voltmeter. The overrange circuits are usually sophisticated enough that they are not only able to handle the two overranges (indicated in the previous explanations) but also to handle a third overrange, permitting the normal range of the voltmeter to be 0 to 1999.

This means a first overrange when the unknown signal is being measured; a second overrange when the reference voltage cannot reduce the integrator to zero in the same time allowed the unknown voltage; and a third overrange if the reference voltage requires twice as long to reduce the integrator to zero as did the unknown voltage to charge the integrator.

A third overrange with a simultaneous detection of zero output from the integrator would indicate an unknown voltage twice the reference voltage. Some voltmeters have provisions permitting even greater overrange capability. Display ranges of 2999 or 3999 are available. Such voltmeters simply make a tradeoff between the voltage swing of the integrator for 0-to-1.000 inputs and the overrange voltage swing which must be permitted for extensive overrange capability.

Polarity. The previously described dual- and single-slope converters have been *monopolar*. That is, they only work with unknown signals of a single polarity. Most digital voltmeters either indicate the necessity for reversing the polarity of the unknown signal to display a proper reading, or they reverse polarity automatically and indicate the proper polarity of the unknown signal with a ± indicator. Single-slope converters and some dual-slope converters simply employ an additional comparator on the input (or at the output of the integrator) which detects an improper polarity input and indicates this on a display. Automatic polarity selection with the dual-slope converter is usually made by changing the polarity of the reference voltage. Signals directing the polarity reference source of the converter are generated by comparators at the output of the integrator.

Offset. To effectively measure unknown signals at or close to 0V, an *offset voltage* is added to the converter. For example, a 1V full-scale converter might actually measure signals from 0.1V to 1.1V at the converter. Thus the dual-slope integrator would never have to integrate an input signal of less than 0.1V. Some dual-slope converters have been developed with an additional cycle in the measurement scheme. This additional cycle is completed with zero input. As the input during this cycle is known to be zero, any reading present at the completion of the cycle can be used to correct zero-offset circuits. Such an additional cycle insures maintenance of a true zero condition.

There are many other variations of the dual-slope converter. One of the first efforts was to remove the switch from the input signal entirely, as switches in the input line present a design problem. Mechanical ones, such as reed relays, have a short lifespan, and long-life semiconductor switches have serious temperature problems.

Voltage-to-Frequency Converter

As the name implies, this converter performs conversion by producing a frequency directly proportional to a voltage. A digital frequency counter may be used for the counting and display circuits.

The V–F converter is a wide-range, highly linear voltage-controlled oscillator (VCO). A typical dynamic range for the VCO might be 10 Hz to 100 kHz. Figure 2-6 is a block diagram of a converter designed to cover such a range linearly. The basic blocks consist of an integrator; a wide-range (but not linear) VCO; an oscillator of known frequency, which serves as a time reference; a logic circuit, which generates a pulse of precisely known width for each cycle of the VCO; and a switched voltage reference source, which upon command from the logic circuit, generates a pulse of precisely known amplitude for the duration of the logic signal.

Operation of the converter is based on the assumption that the polarity of the unknown voltage is the opposite of that of the voltage reference. When a negative unknown voltage is applied to the input, the output voltage of the integrator begins to increase. As the output voltage of the integrator increases, the frequency generated by the internal VCO begins to increase. As the VCO begins to generate signals, the logic circuit begins to generate switching signals. These signals produce pulses at the second input of the integrator. As these pulses are produced by the voltage reference source and logic circuit, they are of precisely controlled amplitude and width.

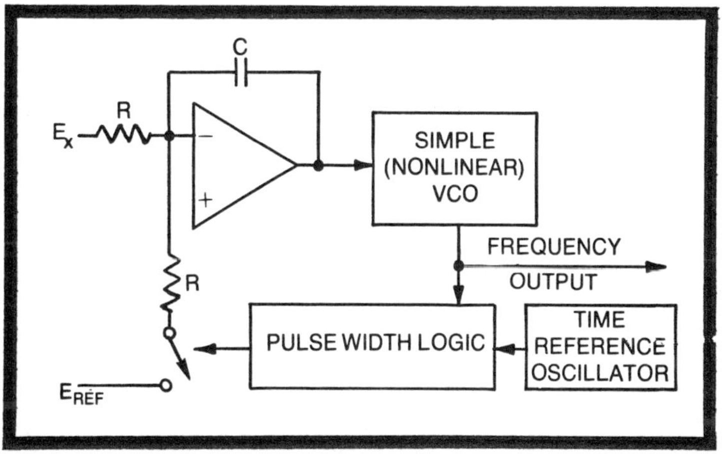

Fig. 2-6. A block diagram of the "glug" type of voltage-to-frequency converter. This converter balances when the energy contained in the precise pulses (glugs) generated by the pulse width logic and in the reference signal E_{REF} is exactly equal and opposite to the energy contained in the unknown signal E_X. The amount of balancing energy is regulated by controlling the frequency of the pulses.

Each pulse therefore represents a well defined amount of energy.

The output of the integrator is a function of the average of the total energy at the input. So long as the average of the energy contained in the pulses is less than that contained in the unknown voltage, the output voltage of the integrator continues to rise. If the total energy contained in the pulses is greater than that contained in the unknown voltage, the output voltage of the integrator falls. And, of course, if the energy of the pulses is exactly equal to the energy in the unknown signal, then the output voltage of the integrator remains at the voltage which causes the VCO to generate pulses at this rate.

Calibration of this converter consists of adjusting either the width or the height of the pulse so the generated frequency is directly related to the unknown voltage. The VCO frequency is then directly proportional to the voltage applied to the input of the integrator and therefore may be measured with a digital frequency meter.

One advantage of this form of analog-to-digital conversion is its ability to reject noise. First, the integrator itself provides a sizable amount of filtering; it acts as a low-pass filter. Second, the gate interval (time) of the digital frequency meter may be made equal to an integral multiple of the period of an interfering signal so that any noise averages out. In other words, the frequency meter itself acts as an integrator. The normal-mode rejection to line-frequency interference is exceptionally high.

A second advantage of this type of converter is again due to the integrating feature. In a true integrator the total number of output pulses represents the area contained under the curve of a varying input signal over the gate interval. If, for example, the area under a curve must be measured, one common method of doing this is to plot the variable either by hand or on a strip chart recorder and then count the number of squares contained under the curve of a varying input signal over the gate interval. If for example, the area under a curve must be measured, one common method of doing this is to plot the variable either by hand or on a strip chart recorder and then counter the number of squares contained under the curve. A second method is to use an integrating voltmeter. The V–F converter fits this application well. The only change from the voltmeter described earlier is that the gate interval of the digital frequency counter is now controlled by the user. To make the measurements, the user counts pulses from the V–F converter over the desired period. The displayed number on the digital counter represents the area under the curve.

A third reason for the popularity of this converter is its adaptability to a digital counting instrument. There are no requirements for special overrange circuits. For a time, the V–F converter also represented the ultimate in accuracy; however, the dual-slope converter has taken this lead at the moment.

DMM MEASUREMENTS

The basic analog-to-digital (A–D) converter is not a complete DMM (digital multimeter) any more than a simple DC analog-meter movement is a VOM. Other circuits must be added to provide the functions of ranging, AC-to-DC conversion, current measurement, and resistance measurement. Many of these circuits are like those used with the analog multimeter, but a few of them take on special aspects when they are applied to the DMM.

DC Voltage Measurement

Measurement of DC voltage is relatively simple if an A–D converter with an autopolarity or polarity indication is available. Figure 2-7 shows the simple addition of a precision voltage divider and a high-input-impedance, unity-gain amplifier. The precision voltage divider reduces high voltages so they match the allowable input range of the converter. The input amplifier must be of sufficiently high input impedance so as not to load the divider; it must have low-noise, high-gain stability, and low drift if this is not taken care of by an autozero circuit.

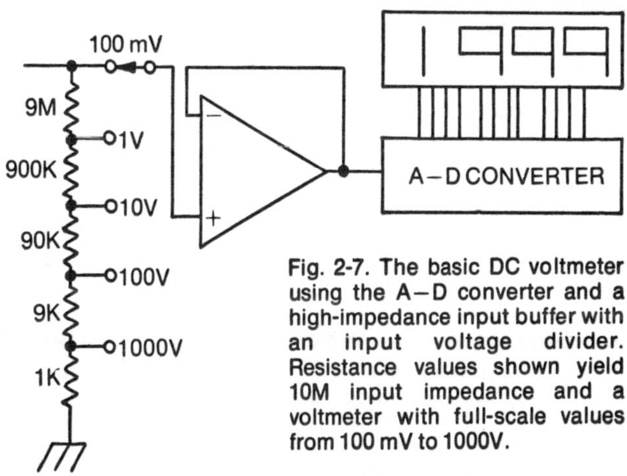

Fig. 2-7. The basic DC voltmeter using the A–D converter and a high-impedance input buffer with an input voltage divider. Resistance values shown yield 10M input impedance and a voltmeter with full-scale values from 100 mV to 1000V.

The voltage divider presents an input impedance of either 10M or 1M. The voltage divider must be of high precision. This can be achieved with precision resistors, hybrid circuits (frequently employed when the divider accuracy exceeds 0.25% or so), or a trimmed voltage divider when the utmost accuracy is desired. The basic voltage range of the converter dictates the minimum range of the DMM unless an amplifier with gain is placed before the converter. Most converters have a sensitivity of either 1V or 100 mV.

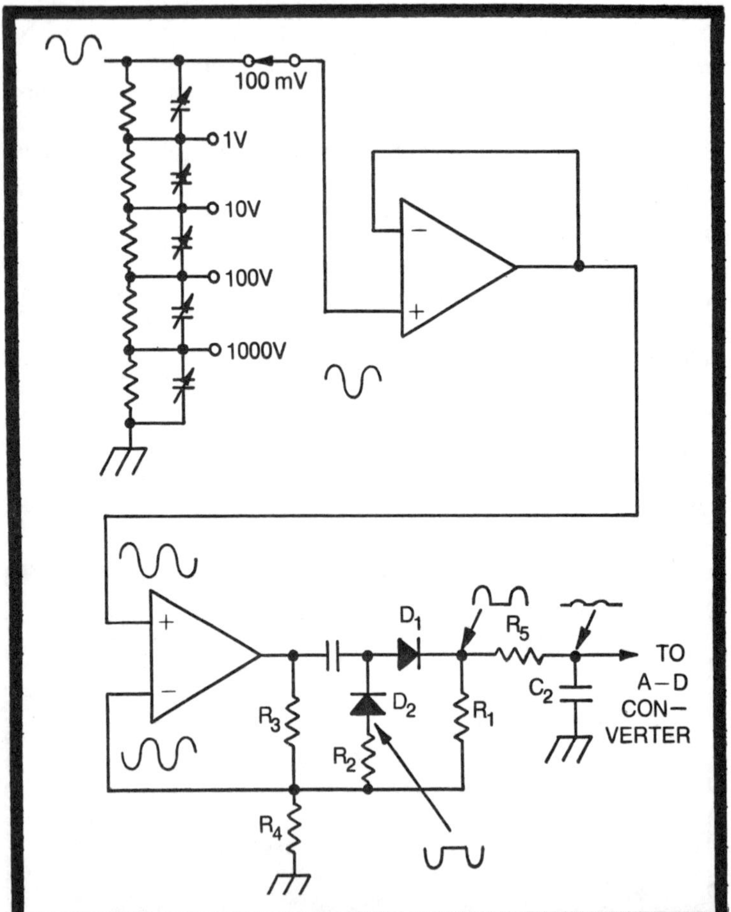

Fig. 2-8. The AC input circuits for a DMM. The same voltage divider is used as in Fig. 2-7, but compensating capacitors have been added to extend operation into the 10–50 kHz frequency range. The operational rectifier converts the input sine wave to half-wave-rectified DC signals. The operation rectifier is utilized with DMMs to eliminate the low-voltage conversion errors which accompany the simple AC rectifiers used with analog meters.

AC Voltage Measurements

Measurements of AC voltage start with the voltage divider. Many DMMs use the same voltage divider for the AC and DC measurements. AC use requires the divider to be compensated for any stray capacitances shunting the divider string. The capacitive division ratio must be the same as the resistive ratio, or a different division becomes apparent as the measurement frequency is increased. Such a compensated divider is shown in Fig. 2-8.

Also shown in Fig. 2-8 is the buffer amplifier and an example of an AC-to-DC converter. This circuit is called the *operational rectifier*. The AC signal is fed to the operational amplifier. On positive half-cycles, feedback to the minus (−) input of the operational amplifier is made through D_1 and R_1. The gain is determined by the ratio of R_1 to R_4. The forward voltage drop of the diode is effectively eliminated by the high open-loop gain of the operational amplifier. The high gain allows the output to change rapidly, driving the diode into conduction. Once the diode conducts, the output amplitude is controlled by R_1 and R_4. The output signal from the rectifier is filtered by R_5 and C_2 to provide a steady DC signal from the A–D converter. On negative half-cycles the feedback path is provided by D_2 and R_2; however, no voltage is contributed to the output. Resistor R_3 provides DC stability during no-signal conditions. The bandwidth of this converter is dependent on the bandwidth of the operational amplifier.

Current Measurements

Current measurements in the DMM are usually made by the same technique as in analog meters—by shunts. The DMM has an advantage over the simple VOM in that the shunts often have a lower voltage drop, permitted by a voltmeter of 100 mV full scale. Shunts are used for both AC and DC measurements. As with many other sections of the DMM, the accuracy requirements for some of the components becomes much greater than required with a VOM. In the analog multimeter a 1% current shunt may be quite adequate. With a DMM, the shunt tolerance may need to be nearer 0.1%.

An alternative to direct-current measurements by shunts is shown in Fig. 2-9. This technique applies the unknown current to the summing junction of an operational amplifier. A high-gain operational amplifier with negative feedback keeps the summing junction voltage near zero. The amplifier output voltage through the feedback resistors generates a second current at the input summing junction equal and opposite to the current at the input. Given decade-value feedback

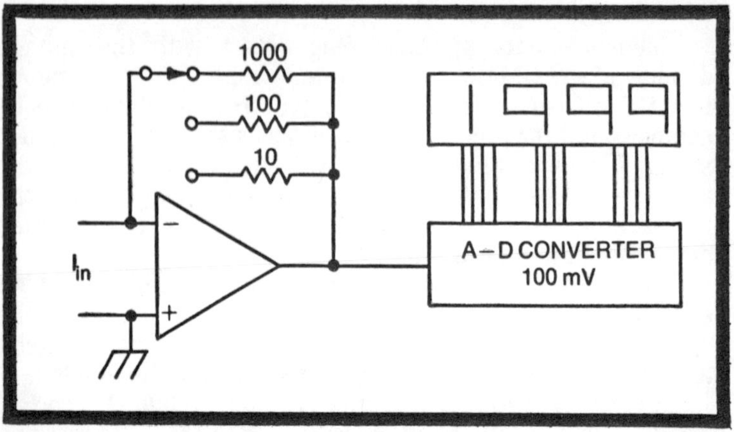

Fig. 2-9. An operational current-to-voltage converter for direct-current measurements with a DMM. The operational current-to-voltage converter provides a nearly zero voltage drop across the input terminals. Resistances shown would yield full-scale values of 100 μA (for 1000 ohms), 1 mA (100 ohms), and 10 mA (10 ohms) for a converter with 100 mV basic sensitivity.

resistors, the output voltage is directly proportional to the current at the summing junction. With this form of current measurement no voltage drop is required to make the measurement, as is required with shunts. There are two limitations to this method of measurement. First, the technique is applicable to DC measurements only. Second, the instrument must be able to generate a current equal and opposite to the one being measured.

Resistance Measurements

Resistance measurements take a different form in the DMM than in the conventional analog multimeter. The analog meter usually uses an internal voltage source applied to an internal resistance in series with the external unknown resistance. The meter has a specially calibrated nonlinear scale which converts the voltage across the unknown resistance to a resistance reading. Nonlinear scales are very difficult to achieve with a DMM; therefore, a different technique is employed. A constant-current source generates the test signal to be passed through the unknown resistance. A constant current of a known value develops a voltage across the unknown resistance directly proportional to the unknown resistance. The voltage is measured and displayed as resistance. Such a measurement system is shown in Fig. 2-10. The circuit used for the constant-current source may be either discrete (Fig. 2-11A) or integrated (Fig. 2-11B).

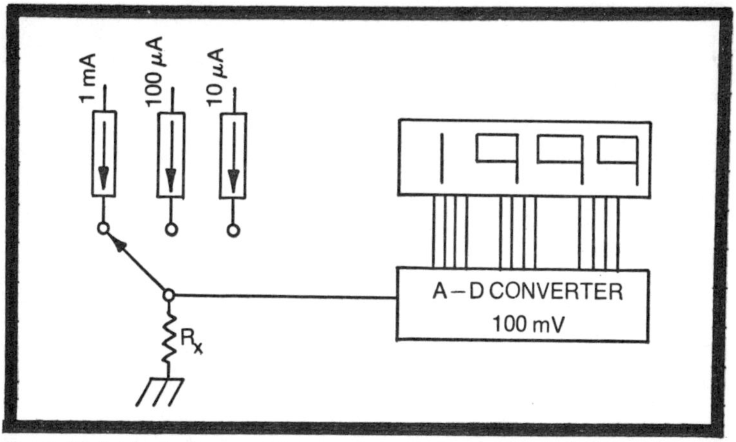

Fig. 2-10. A block diagram of the resistance-measuring technique used in the DMM. Using a 100 mV converter, the 1 mA, 100 μA, and the 10 μA current sources give full-scale resistance ranges of 100, 1000, and 10,000 ohms, respectively. Even if the 100 mV range of the meter is used, diodes and other semiconductor junctions in the circuit being analyzed may be driven into conduction, as most current sources used in this application have open-circuit voltages 10 to 100 times greater than the converter range.

Constant-current ohmmeters have a few special characteristics. First, the open-circuit voltage is high, often a few volts. Unless specially designed, these ohmmeters forward-bias semiconductor junctions. Second, as the current ranges must be the same number of decades as the resistance ranges, the high-resistance ranges use low currents. A small current charging a capacitance takes appreciable time, so the high-resistance ranges tend to be slow responding. These types of meters are used most often at upper limits of about 20M.

Autoranging

One of the best features of the DMM is almost hands-free operation. Even greater ease of operation is given with the autoranging DMM. Special circuitry is added to the conventional DMM to sequence the instrument until the display is all but overranged. The techniques for autoranging are varied and involve both electronic and electromechanical techniques.

SPECIFICATIONS

There might seem to be little to specify on the digital multimeter except the functions, ranges, and accuracies. But the DMM has extensive specifications. Many of the variations from DMM to DMM are subtly specified and can make all the

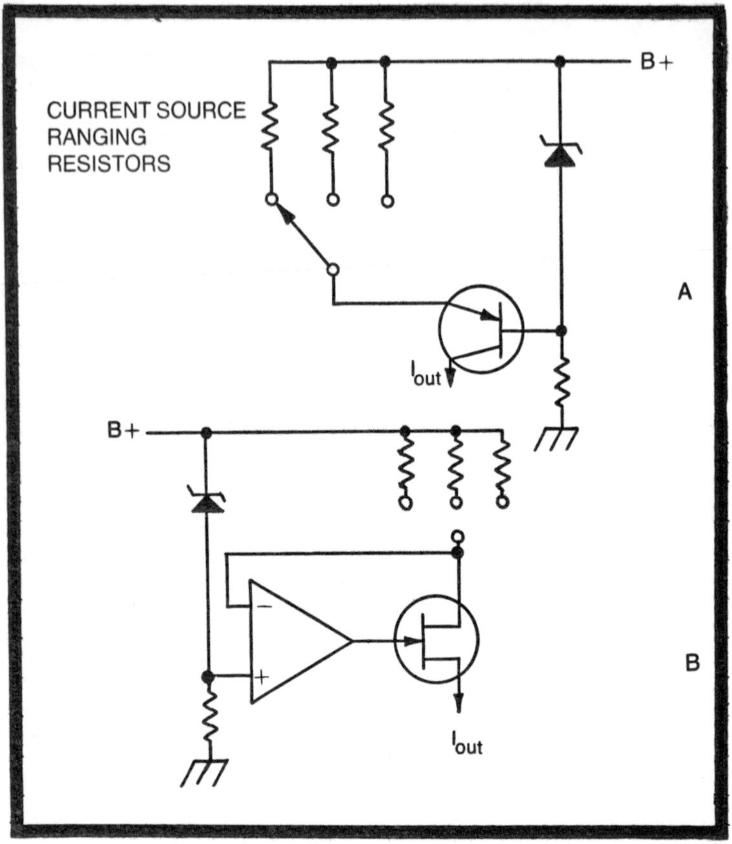

Fig. 2-11. Two forms of constant-current sources. (A) A positive constant-current source using a PNP transistor. The base—emitter junction of a second transistor may be placed in series with the zener voltage reference to provide temperature compensation. (B) A constant-current source using an N-channel junction FET and an operational amplifier.

difference in the world to the user. This is especially true in field use of the instrument.

DC Voltmeter

Ranges. One of the first questions facing the potential buyer is defining full scale on a particular multimeter. This is specified in one of two ways: either as 1, 10, or a multiple of 10, with usable overrange capability specified as a percentage (typically 100%); or as the maximum possible reading. For example, a DMM may be specified as having a 1V full scale with 100% overrange, thus indicating useful operation to 2V, or it may be simply specified as having a 2V full scale. These ranges are further limited, as the full indicated capability of the meter may not be useful on the highest voltage range. For example, a DMM with a 1999 full scale may not be able to read over 1000V DC (even lower on AC). This is usually because of voltage breakdown ratings of internal components.

Accuracies. Accuracy specifications differ with the manufacturer, as well as with the degree of accuracy. A meter with very high accuracy will have more sophisticated accuracy specifications than a meter with limited accuracy. Simple accuracy specificatons are given as ±% of full scale, ± 1 digit. The "±1 digit" portion of the specification relates to an error in the digital counting circuits, the "±% of full scale" includes ranging and A–D conversion errors.

One of the most sophisticated specifications is ±% of reading, ±% of full scale, ±1 digit. Such a specification is usually confined to instruments in the 0.01%–0.05% class. In the 0.05%–0.5% class the "±% of reading ±1 digit" specification is quite popular.

An additional specification may qualify the accuracy of the instrument at temperatures other than 25°C. Temperature specifications are of two forms: (1) a temperature coefficient—percentage per degree Celsius (centigrade)—with which the user may calculate the exact deviation from the 25° specification, knowing the ambient temperature; or a complete temperature range, such as 15°C to 35°C.

Other limitations may be placed on the accuracy of the instrument. These include the effects of line voltage variations, humidity, altitude, time, and more. These limitations are of little interest to the person making general use of the multimeter; however, as these are very portable instruments, some manufacturers, not knowing where they will be used, issue all-encompassing specifications. One thing you can be sure of—the more inclusive the specifications, the higher the cost of the instrument.

Input Impedance. Most DMMs have a 10M DC input impedance, a few have a 1M input impedance. Input impedance may have a tolerance specified. This is important when using the meter with an external multiplier resistor. Some voltmeters offer very high input impedance on the lowest DC input ranges. Input impedances on such DMMs may be in the 100M–1000M range.

Response Time. The response time consists of two factors: first, the basic cycle rate of the A–D converter; second, the time required to charge capacitances found in the input circuits. This latter may be quite long if there is input filtering. Response time is the number of seconds required for the instrument to settle to its rated accuracy. In lieu of response time, some manufacturers simply give the number of conversions per second.

Protection. Protection specifications indicate the amount of line frequency AC overload each range will tolerate without damage. This is especially important when using the instrument in an industrial or other application where accidental contact with 120V or 240V AC is quite possible.

Normal-Mode Rejection Ratio. The NMRR indicates the amplitude of AC (usually line frequency) interfering signal impressed on the DC being measured which will affect the least significant digit (see Fig. 2-12). The ratio of the interfering signal to the voltage represented by the least significant digit is usually expressed in decibels. For example, an instrument reading 100.0 mV DC is specified to have 60 dB NMRR. The least significant digit (100 μV) indicates that 100 mV will not affect the reading in the least significant digit; any signal greater than 100 mV may. The NMRR depends upon the instrument timing and may have to be adjusted for change in power line frequencies (e.g., 50, 60, and 400 Hz).

Common-Mode Rejection Ratio. The CMRR specifies the instrument's ability to reject signals applied between earth ground and a point common to the "high" and "low" input terminals of the instrument. There is no CMRR specification if the instrument has its low terminal common to earth ground. Figure 1-13 indicates the common method of measuring CMRR. The 1K resistor in series with the low terminal is generally included with any CMRR specification. This resistor represents a typical source resistance of DC signals under actual measurement conditions. Current flowing in the common-mode path flows through the resistor. The voltage generated across the 1K resistor is converted to a

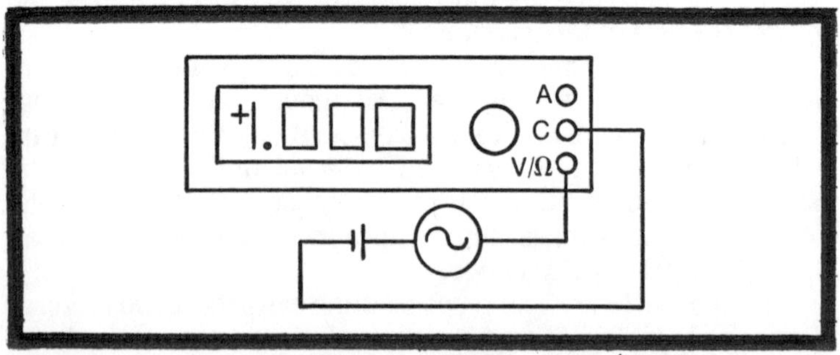

Fig. 2-12. The test configuration to measure normal-mode rejection ratio (NMRR). The amplitude of the AC series signal is increased until the least significant digit of the display is changed. The NMRR in decibels will be 20 log (E_{ac}/E_{lsd}), where E_{ac} is the value of the AC signal and E_{lsd} is the DC value of a 1-digit change in the instrument's least significant digit.

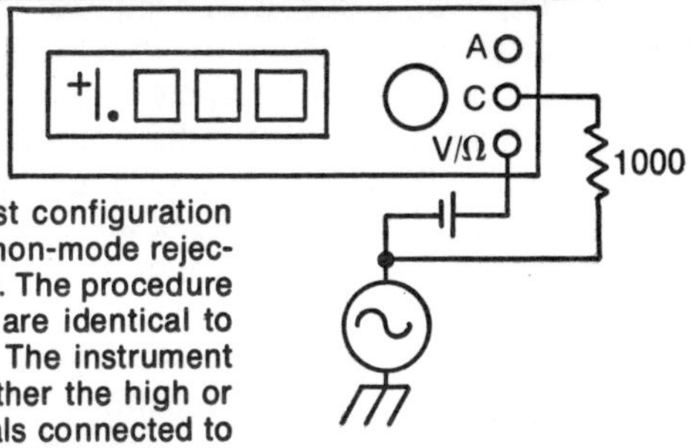

Fig. 2-13. The test configuration to measure common-mode rejection ratio (CMRR). The procedure and calculations are identical to those for NMRR. The instrument must not have either the high or low input terminals connected to earth ground. The CMRR may be virtually impossible to measure in battery operation.

normal-mode signal, which is rejected by the instrument's NMRR. Occasionally CMRR is given less the NMRR. Generally the CMRR *includes* the NMRR. As with NMRR, the CMRR is given at power line frequencies. The CMRR will worsen with increasing frequency.

DC Common-Mode Rejection Ratio. The DC CMRR, or the *floating capability* of an instrument, is often limited by the breakdown voltages of the input circuitry. This specification indicates the greatest DC potential the low terminal of the voltmeter may have above earth ground.

AC Voltmeter

Ranges. Range specifications are of the same nature as those given for the DC voltmeter. The high-voltage range may have an upper voltage limit considerably less than expected from a front-panel reading; 750V is common.

Accuracies. AC voltmeter accuracies are generally given in a manner common to the DC voltage accuracies. However, accuracies are normally only for measurements of sinusodial signals with less than a specified amount of harmonic distortion (usually 0.5%). AC-to-DC converters, which are normally average or peak responding but RMS reading, require this limitation; if waveforms other than sinusodial are measured, the accuracy specification no longer holds. This is not true if the instrument employs a true-RMS converter. True-RMS converters are not common and are very expensive. The normal range for AC accuracies is 0.5%–1% for the average- or peak-responding RMS-calibrated instruments.

Most AC voltmeters will have a frequency response specification. Frequency response indicates the instrument's ability to measure high-frequency signals, also the expected inaccuracies over a specified frequency range. The limits of AC frequency response are normally from 20 Hz to 10 kHz or 50 kHz, depending on the instrument.

Input Impedance. The input impedance specification of the DMM should include not only the resistive value to be expected (usually 1M or 10M) but also the value of capacitance found between the input terminals. This capacitance is generally about 100 pF.

Response Time. AC voltmeter response time includes all the time specified in the DC voltmeter as well as the response time of the AC converter. AC response time may be 6–10 times greater than for the DC voltmeter.

Input Protection. Input protection indicates the amount of voltage overload which may be applied to any range without damage. A separate DC limit may be indicated to cover input-coupling-capacitor breakdown. Overloads from sources outside the specified frequency range of the instrument may not have as great a protection range.

Common-Mode Rejection Ratio. This is defined and measured in the same manner as for the DC voltmeter.

Noise. Some of the very good voltmeters indicate the RMS value of noise contributed by the converter, the input amplifier, and any other source within the instrument. A noise specification is only required on very high resolution, sensitive instruments.

Ammeters

Ranges. Ammeter ranges are given as full-scale readings and may include an overrange specification. A number of instruments do not have extensive ammeter ranges; other meters commonly extend to 1A full scale. Some instruments have DC capabilities only. Ammeter ranges vary extensively, so these specifications must be carefully read. All ammeter ranges have full overrange capability; therefore, a 1A meter usually gives 2A capability.

Accuracies. Ammeter accuracies will be slightly lower than voltmeter accuracies, as the accuracy of the shunt must be included. Accuracy of the ammeter may be further degraded by high-current shunts.

Voltage Drop. When inserted into circuits, the ammeter shunt causes a maximum voltage drop when measuring full-scale currents somewhat larger than the full-scale value of the lowest voltmeter range of the instrument. This may be as much as 10% or 20% higher than the voltage range in order to cover resistance in series with the shunt, especially on the highest current ranges, where the shunt value is usually 0.1

ohm. On very low current ranges the shunt resistance is relatively high. For example, a 200 pA range on a 200 mV meter will have a 1000-ohm shunt.

Protection. Most DMMs have a fuse in series with the ammeter shunts that opens if the maximum current is exceeded. It is wise to note fuse types. A few DMMs use very unusual fuses, and keeping a few spares on hand may save time and trouble.

Response time. Response time of the ammeter should be similar to that of the corresponding voltmeter (AC or DC).

Ohmmeter

Ranges. The lowest ohmmeter range on most DMMs is higher than expected. Usually the first range is 100 ohms. A 100-ohm range will give 0.1-to-1-ohm resolution. The upper limit of the ohmmeters found in DMMs is either 1M or 10M. Ten megohms is the more desirable. Ohmmeter ranges are found in decade steps between 100M and 16M. All ranges have full overrange capability, so a 10M meter normally gives 20M measurement capability.

Accuracy. Accuracy of the ohmmeter measurements is related to the accuracy of the DC voltmeter and the precision of the constant-current sources. The accuracy specification may be somewhat reduced for measurements on the uppermost range, but for most ranges the error is no greater than twice the DC error.

Measurement Currents. Some DMM manufacturers only specify the current applied by each resistance range to the unknown resistance, while others specify both the current and the maximum open-circuit voltage applied to the circuit being tested. Some DMMs have special low-voltage ranges so that semiconductor junctions are not forward-biased.

Response Times. Resistance measurements normally have a response time close to that of the DC voltmeter. The uppermost range, however, may have a response time considerably slower than that of the other ranges.

Protection. Protection of the ohmmeter is important, as the constant-current generator is easily damaged if a high external voltage reaches it. Protection may differ for AC and DC and may vary to some extent with the resistance range being protected. One of the most desirable voltages to have protection against is the power-line voltage. Accidental contact with this high potential is not at all uncommon, and a DMM without 120V AC ohmmeter protection is vulnerable to extensive damage. Many DMM ohmmeter circuits employ a very small fuse as part of the protection. Once again, this fuse may be uncommon, and obtaining a few spares is wise.

GENERAL FEATURES

Often the features are what make one instrument stand out from another. The potential DMM buyer should have a good knowledge of an instrument's features and how he can use them. Frequently the purchaser needs to make a tradeoff between features, so careful consideration of one feature against another is in order so as to gain maximum use from the DMM investment.

Digits of Readout

One of the strange terms used by DMM manufacturers is *half-digit*. It is perfectly clear what a 3-digit DMM is, and what a 4- or 2-digit "machine" (as they are sometimes referred to) is; but a 3½-digit meter sounds like something run through a saw. The term *half-digit* has been coined to indicate a DMM with 100% overrange capability. This is also referred to as *1999* capability. In a similar manner the term *3¾-digit* refers to a DMM with *3999* capability. Whenever terms such as these are used, it is wise to inspect the specifications very carefully to determine their exact meaning. Good specifications give a numerical upper limit to each of the instrument's ranges and also specify overrange capability.

The *resolution* of the instrument is directly limited by the number of digits in the display. A 3½-digit DMM has a resolution of 1 part in 2000, or 0.05%. If, for example, the 1V range has a full-scale value of 1999 mV, the resolution is 1 mV. Four-and-one-half-digit machines have resolutions of 1 part in 20,000, or 0.005%; but unless the noise is very low, this resolution may not be usable, as there may be instability in the last few digits at all times. The 2½-digit instrument has a resolution of 1 part in 200, or 0.5%.

A 2½-digit or a 3-digit DMM should be considered as a replacement for a good analog multimeter, so far as accuracy and resolution are concerned. Accuracy generally lies between 0.5% and 1.5%, with 0.5% resolution or more. The 3½- and 4-digit machines generally have accuracies between 0.5% and 0.05%, and 0.05% resolution. Such resolution and accuracy generally suffice for even the most exacting service work and home experimentation. A capability of 4½-digits or more generally indicates an accuracy of 0.05% or better, with 0.005% resolution; this is considered laboratory instrumentation.

Power Sources. Most DMMs today offer battery as well as power-line operation. A few, either the very exotic or the very inexpensive, do not. Frequently the battery feature is an option at considerable extra cost, especially if the batteries are rechargeable. When considering battery operation, note the type of cells used. Replacement of an odd-sized cell may be both difficult and expensive. Rechargeable cells which are physically and electrically interchangeable with zinc–carbon or alkaline cells have an advantage. Temporary substitution permits operation in a portable situation even if the nicads were not charged the night before. All cells have a finite life expectancy, and today none are particularly cheap. Therefore, if there is no need for a portable instrument, battery operation may not be worth the price. Battery operation is not confined to field use; even on the bench, battery operation may be an advantage in making a voltage measurement with the DMM floating at a potential above the allowable common-mode voltage of the instrument.

For truly portable operation the operating time from a full charge is an important specification. For extremely constant use an operating time of 8 hours may be needed from a single overnight charge; however, if the DMM is to be operated intermittently during the course of a working day and is to be kept on a charger overnight, an operating time of 4–6 hours will be quite satisfactory and probably cheaper.

Status Indication. Digital multimeters have many modes of operation. With either an autoranging instrument or an instrument being operated at some distance from the user, some form of status indication is convenient. Status indication displays the DMM function and the range being used. Usually it takes the form of lighted indicators in the display window. Overrange decimal point and polarity are the most frequently included status indicators. Be certain these three indications are easily understood. Blinking or blanking of the display is frequently used to indicate an overrange condition. Illuminated + and – symbols most frequently indicate the polarity of the DC measurements.

Sample Rate. The sample rate specification indicates the number of conversions in one second. Commonly this figure is about three to five per second. With 7-segment displays, sample rates in excess of five per second may create readings which are difficult to read.

Warmup Time. Many instruments specify a period of time required before the instrument is within its specifications. Quick warmup may cost extra, but if high accuracy and rapid portability are requirements, as in certain types of service work, this may be a feature worth paying for.

Operating-Temperature Range. The accuracy specifications of a DMM have a temperature dependency. This usually is specified in one of three ways. First, the DMM may be given a temperature range over which it may be operated within its published specifications. Second, the DMM may be given an accuracy specification at 25°C and a derating figure for temperatures other than 25°C. Third, the permissible operating-termperature range of an instrument regardless of accuracy may be given. This is important. Work done in cold climates may find the instrument kept in an unheated portion of a service truck. The instrument with an operating temperature range of 0°C and above may well not operate on a moderately cold winter day.

Size and Weight. The physical characteristics of a DMM make it portable. They also contribute to price and complexity. Again, keep the intended application in mind to make the best cost–value tradeoffs.

Displays. The light-emitting diode (LED) is one of the most popular displays used with DMMs. Other displays in use are the 10-character neon display (Nixie), the 7-segment neon display, the fluorescent display, and the liquid crystal display (LCD). LEDs are popular because of their good brightness, excellent contrast, and low cost. Neon displays, both 10-character and 7-segment, have the highest brightness but at a slight increase in cost, combined with the requirement for a high-voltage power supply. Neons also tend to generate some slight RF noise. Fluorescent displays have never been too popular, although they are generally lower in power consumption than LEDs or neons. Fluorescents are subject to interference from static electricity and they have poor contrast. Extremely low-power operation makes LCDs popular. They also have a potentially low cost. They are, however, the displays with the lowest contrast ratio. Certain types of LCDs don't wash out in direct sunlight, but most LCDs will freeze at moderate temperatures and become completely useless. The life expectancy of LCDs is one-hundredth (or less) that of the neon or LED displays.

When considering displays, size must be given some thought. DMM displays will run from 0.1 in. high to 0.6 in. or more. Often the user is never more than the length of the test leads away from his DMM. In such cases small displays are no hindrance and permit a smaller, more portable design. On the other hand, if the contemplated use is more laboratory oriented, readings may be required at a greater distance, and larger displays may be necessary. Again, a consideration of the instrument usage must be made.

APPLICATIONS AND SOURCES OF ERROR

An applications section on DMMs seems almost superfluous, since the instrument simply measures current, voltage, and resistance. While this is true, there are a few special situations in which the DMM is used that are worth discussion.

Probably one's first impression after using the DMM is a feeling of "How did I ever get along without this instrument?" This attitude results from the increased convenience. A 3½-digit, autopolarity machine rarely needs range changes when working with most circuits. One range, such as the 20V range, gives all the resolution required. Without having to reach for the POLARITY switch, there is nothing left to do but take measurements.

For example, a 3½-digit machine on the 20V range has a full-scale reading of 19.99V. Most power supplies of modern analog circuits can easily be checked to the nearest 10 mV, and the base—emitter voltage drop of transistors still checks to two significant figures (again the nearest 10 mV). Such measurements give more than adequate resolution. The 10 mV resolution is in the range of the voltage changes with temperature for a forward-biased diode.

Needless to say, the DMM is not without its pitfalls. Erroneous actions based on DMM readings, assuming more accuracy than exists, or problems of readings with too much resolution are frequent. For example, instructions in one Heath oscilloscope kit manual direct the builder to adjust a control until voltage on the collector of each of two deflection transistors is equal, then to adjust another control to set both collectors at 100V. A number of kit builders have found this task frustrating and next to impossible. The reason: A DMM was being used which had far more resolution than called for. Adjustments were being made to the nearest few tenths of a volt which need only have been made within a few volts. Adding a factor of 10 to the settability of a control can make the difference between one that adjusts simply and one that is difficult to adjust.

In a similar case an error is often made when a voltage is not exactly the value that is expected. For example, the output voltage of a regulated power supply is specified to be 15V plus or minus 0.5% (75 mV). On inspection with a DMM, the voltage is found to be low, 14.90V. Here is a case for concern if circuits powered by this supply are not meeting specifications. However, it is well to do a little thinking before immediately repairing or adjusting the supply. First, is the problem the circuit is showing likely to be caused by a power supply 25 mV below specs? The likelihood is it's not. Therefore the real problem must be determined first. Once the major problem is discovered and repaired, all specifications can be checked. If all specs are fine, it may be that readjusting the power supply to its correct voltage will do no more than throw off the calibration. The product may well have been initially calibrated and adjusted with the low power supply.

The caution implied in both examples above may be stated: Don't overuse your DMM. Most of the electronics we work with is designed with about 10% tolerances. The DMM has high accuracy and resolution, but when these are not needed, learn to disregard them.

Peaking and zeroing are two adjustments common to electronic circuits, especially those employing tuned circuits. Peaking or zeroing with a digital instrument is not easy. The analog meter gives a very good idea of trend, but to use a digital meter for this purpose, the mind must act somewhat like a digital computer. First it must note one reading, then a second. Then it must compare the two readings and determine which is larger. Then and only then can one know if the adjustment is in the right direction.

Often the resolution of the DMM shows the strangest things—some good, some bad. The DMM with 1 mV resolution easily shows voltage drops across a PCB foil. Perhaps this voltage drop is the culprit. Then again, perhaps the circuit will completely "ignore" this minute voltage and so should you. As noted before, the DMM often exposes variations in semiconductor components with temperature; but unless the circuits being analyzed are extremely critical, few if any problems occur from this source.

Circuit loading becomes much more noticeable when the DMM is used. A DMM with 10M input impedance will load a 50K circuit by 0.5%—unnoticeable on the analog voltmeter, but a sizable change on a DMM with 0.05% resolution.

AC measurements are especially susceptible to erroneous readings. Frequently the accuracy of the AC measurement is an order of magnitude (factor of 10) less than the resolution of the instrument. For instance, a reading may be taken to the nearest 10 mV; but the accuracy may be only 100 mV.

Distortion can also cause problems. The total harmonic distortion of the sinusodial signal being measured must be low to insure the rated accuracy. The power lines may easily have a total harmonic distortion of 1%–3%, thus precluding measurements to the full accuracy of the DMM. Remember, the eye can just notice distortion of 3% or more on an oscilloscope, so don't be fooled by a clean looking sine wave.

Not all measurements beyond the accuracy limits of the instrument need be regarded as useless. For an example, suppose a 3½-digit machine with an accuracy of 0.5% is to measure a 2V source, and a compensating component in the circuit is to be adjusted so the output of the circuit does not vary by more than 2 mV. Two millivolts is 0.1% of 2V, or five times smaller than the accuracy of the instrument. If the DMM reads 1.995V, you do not know whether the actual voltage is 1.985V or 2.005V. However, if the voltage varies with changing load from 1.995V to 1.990V, you do know the output is varying by 5 mV. Adjustments can now be made to reduce this variation to 2 mV or less.

At times, a digital instrument introduces noise to the circuits where they are attached. This noise is usually generated by the logic circuits within the DMM. If the noise is examined with an oscilloscope, one finds they have a period coinciding with the sampling rate of the DMM. It is infrequent that such noise causes a problem; however, when working with wide-bandwidth amplifiers of high sensitivity, suspect the DMM if such a problem should occur.

SUMMARY

Certainly the popularity and longevity of the DMM suggest the instrument's being more than a passing fad. Such an instrument has a place in almost every shop, lab, and field service pack. The question now arises, what next? Have the instrument manufacturers done all that can be done with the DMM? The answer to that question is, absolutely not!

Perhaps the next changes in the DMM will not be quite so dramatic as those of the past decades, but there is still room for improvement. The size of the instrument is not reduced to the smallest possible, or desirable; accuracies can be improved, although they are quite good now; display techniques have a long way to go; power dissipation can be lowered to make the battery life equal to shelf life; and considerable experimentation is taking place on the basic physical construction of the instrument. All in all, there is a great deal to watch for in this area of low-cost electronic instruments.

Chapter 3
Digital Frequency Counter

Until recently the digital frequency counter has been classified as one of the more exotic laboratory instruments. Typically, in the past, the digital frequency counter cost over $1000, and relatively few home experimenters had access to them. The introduction of low-cost digital integrated circuits and matching displays has changed this picture. The new prices of under $200 for a kit and not much over $200 for an assembled instrument mean that even the amateur and the home experimenter can now afford this versatile equipment. The uses for this device range from testing a new oscillator design to servicing an electronic organ. To the radio amateur it represents an ability to comply with the requirements of the FCC rules calling for an independent measurement of transmitter frequency.

FUNDAMENTALS

Figure 3-1 is a much-simplified block diagram of the frequency counter. The unknown frequency F_X is supplied to one input of a gate. The control signal allows the unknown frequency to pass through the gate for a known period of time, such as 1 sec. As the unknown signal passes through, it is accumulated by a series of decade counters, each having its own decoding and display facilities. At the end of 1 sec the number accumulated in the decade counters is the counts per second, or the frequency in hertz.

So long as decimal point placement is taken care of, the unit of time can be decade multiples or submultiples of 1 sec. If

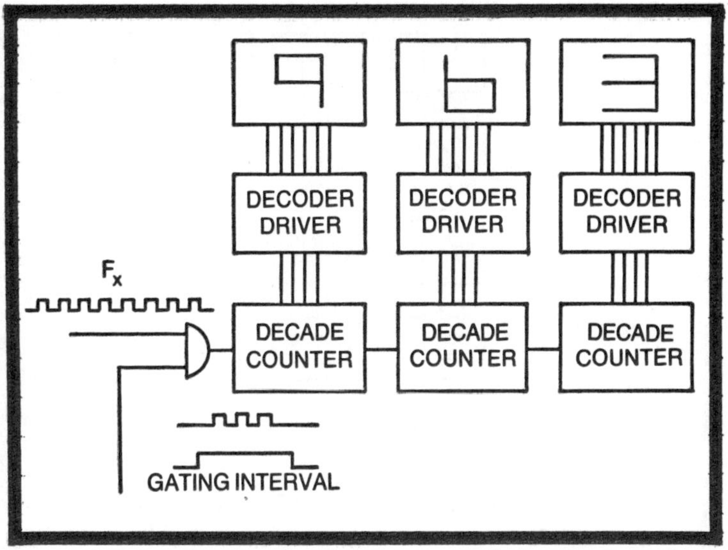

Fig. 3-1. A gate, a few decades of counting and display circuits, and precise control of the open time of the gate are all that are required to create a simple frequency counter.

the number of events occurring over 1 sec is counted, the result displayed is in hertz. If the known interval is 1 msec, the result will be in kilohertz.

Figure 3-2 is a block diagram of a complete frequency counter. In Fig. 3-2 the simple gate and counting mechanism have been supplemented by a crystal oscillator and a series of decade dividers, which are collectively called a *time base*. The time base provides the precisely controlled gating interval. The crystal oscillator provides the precise frequency, and the dividers provide the various decade multiples of time desired. Control logic monitors this gating interval and determines when the display memory should be updated and when the counting decades should be reset; it also supplies the exact time interval to the gate.

The unknown signal does not pass directly into a logic gate, but first goes through shaping and amplifier circuits. Here the input signal is buffered to give a high input impedance and amplified so the basic sensitivity of the instrument is substantially increased. This amplified input signal is also conditioned to be of the proper impedance and voltage level to match the digital logic.

There is also a power supply that delivers the required DC voltages for the various portions of the instrument. The power supply may derive these voltages from the 120V power mains,

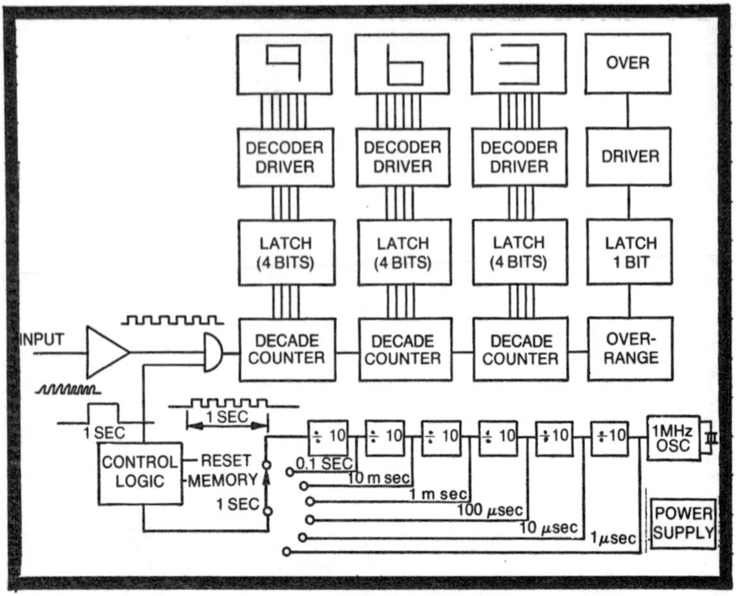

Fig. 3-2. A block diagram of a complete digital frequency meter. The reset pulses are applied to the decade counters after the gate interval is complete and a memory pulse has tranferred the information in the decade counters to the latches. The overrange circuits act as a single bit of counting, storage, and display; but they may only operate once per gate interval.

from an internal battery pack, or from an external DC voltage source, depending on the design.

Higher Frequency Measurements

When the frequency counter is to measure higher frequencies, a number of circuit changes is required. The most straightforward or *direct* approach, of course, is to increase the frequency capability of the input amplifier, the gate, and the first few decades of counting and display circuits. While this seems the obvious way, it is more expensive than other techniques.

Prescaling. Another technique is to insert a prescaler between the input amplifier and gate as shown in Fig. 3-3. In the figure, a divide-by-2 circuit (flip-flop) has been placed between the input amplifier and gate. As the only purpose of this circuit is to divide the unknown frequency by 2, it is a less complex and therefore less expensive method of achieving high-frequency performance than an entire decade counter. The output of the divide-by-2 circuit is one-half the unknown frequency. The gating interval must be extended for twice as

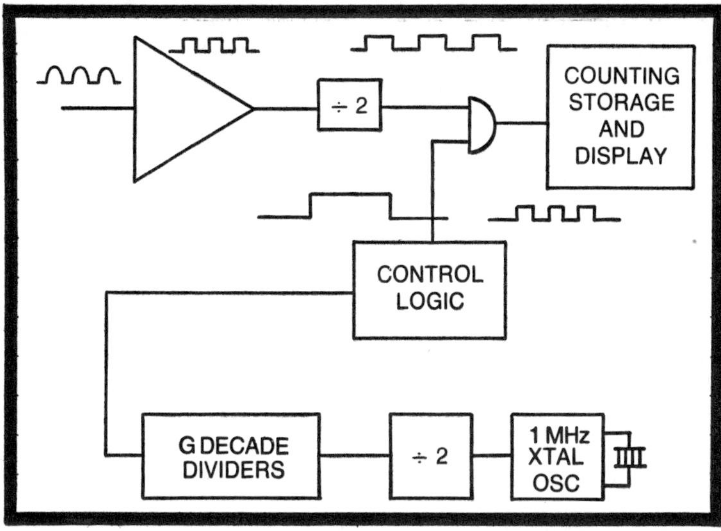

Fig. 3-3. The digital frequency counter modified for extended frequency range by the addition of a ×2 prescaler. Note that a ×2 prescaler has also been added to the time reference line. Prescaling comes in ×2, ×4, ×8, ×10, and ×16 factors. The binary prescaling factors are most common for higher frequency techniques.

much time to again display the proper frequency. To do this, a second flip-flop is inserted between the crystal oscillator and scaler. This doubles the gate interval.

The prescaling technique is quite common at frequencies approaching the state-of-the-art limit for logic circuits. Readily available decade counters extend to a little over 200 MHz today. Beyond that, prescaling appears to offer the most economical frequency measurements. The disadvantage, of course, is a longer wait to obtain the same resolution. That is, if the divide-by-2 is used, the wait is twice as long; and if a divide-by-10 is used, it is 10 times as long. Generally, such additional measurement time causes the nonsystem user no real inconvenience. The direct-measurement method becomes a requirement when a computer or other such data aquisition device is connected to the frequency meter.

Heterodyning. Figure 3-4 demonstrates an alternate method of extending the frequency range of a digital frequency counter. The unknown high-frequency signal is applied to a wideband input amplifier, which in turn applies this signal to a mixer. Here the unknown frequency is mixed with a signal which will heterodyne the unknown signal to a frequency range which the frequency counter is capable of handling. The example of measuring the output frequency of a 450 MHz FM transmitter with a frequency counter of 100 MHz capability illustrates this. If the 450 MHz is mixed with 400 MHz, one of the resultant signals will be 50 MHz. This is in the range of the frequency counter. By mentally adding 400 to the displayed frequency, the exact frequency of the transmitter will be known. The injection signal for the mixer is generally derived from the time base of the frequency counter. Therefore it is a high-stability crystal-controlled signal. This signal, usually from a 1 or 10 MHz crystal oscillator, is applied to a harmonic generator, whose appropriate harmonic is selected by filtering. Such heterodyne converters are becoing less popular as direct counting and prescaling techniques become more prevalent. It is, however, a good technique to keep in mind for the advanced experimenter, as it is possible to design and build such a converter and greatly extend the range of a limited-range frequency counter.

SPECIFICATIONS

To use the frequency counter properly, the manufacturer's specifications must be understood. What do these various specifications mean, and most important, how do they limit

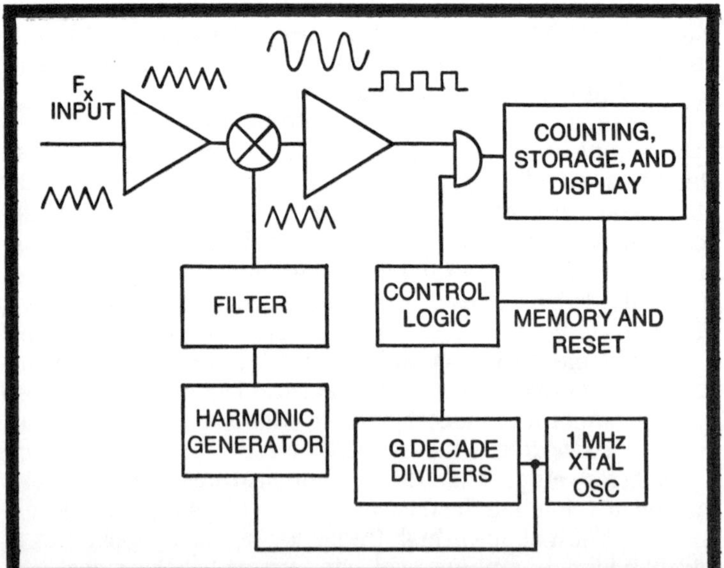

Fig. 3-4. A block diagram of a digital frequency counter equipped to make high-frequency measurements by heterodyning. This technique is normally used on microwave counters, whose price is out of the scope of this book; however, some older units are available on the surplus market which make use of heterodyning.

the measurements which can be made? Specifications are a statement of limitations by the manufacturer and must be treated as such for effective use. The following discussions of some individual specifications will be helpful.

Frequency Range

Generally speaking, most frequency counters are AC coupled, so the usable frequency range of the frequency counter will be specified from some lower limit—such as 1, 5, or 10—to some upper limit, usually the counting limit of the first decade of the counting section. Typical ranges seen on low-cost instuments are between 30 and 200 MHz. The cost of the frequency counter reflects the upper frequency limit to a high degree. If the frequency counter has DC coupling, there will usually be two specifications, such as DC–100 MHz and 20 Hz to 100 MHz. Although DC coupling has no use in a pure frequency counter, on more universal counting instruments AC–DC inputs are very important.

Sensitivity

Sensivitity is a measurement of the lowest amplitude signal that will produce reliable counting within the frequency range. Sensitivity is an important specification; the more sensitive the instrument, the easier it is to couple signals into the input. Typical sensitivities on modern counters are in the area of 10–15 mV and are usually expressed as an RMS voltage. In addition to a guaranteed minimum sensitivity level, manufacturers often publish an input frequency response curve, showing the guaranteed and typical sensitivity versus frequency for the instrument. Such a curve is shown in Fig. 3-5.

Input Impedance

Frequency counter inputs usually fall into one of two classes. They are either 1M shunted by some small capacitance, such as 15 pF; or they are 50-ohm nonreactive inputs, which insures the proper termination of coaxial cables. Generally, low-frequency inputs (that is, those for use below 100 MHz) are of the 1M type, and high-frequency inputs (100 MHz and up) are of the 50-ohm type. This is not an absolute rule, and both types of inputs can be found at higher and lower frequencies. The input impedance specification is important, as it indicates how the circuit under test is being loaded.

Input Protection

There is a maximum level of signal above which the frequency counter will either not function properly or may

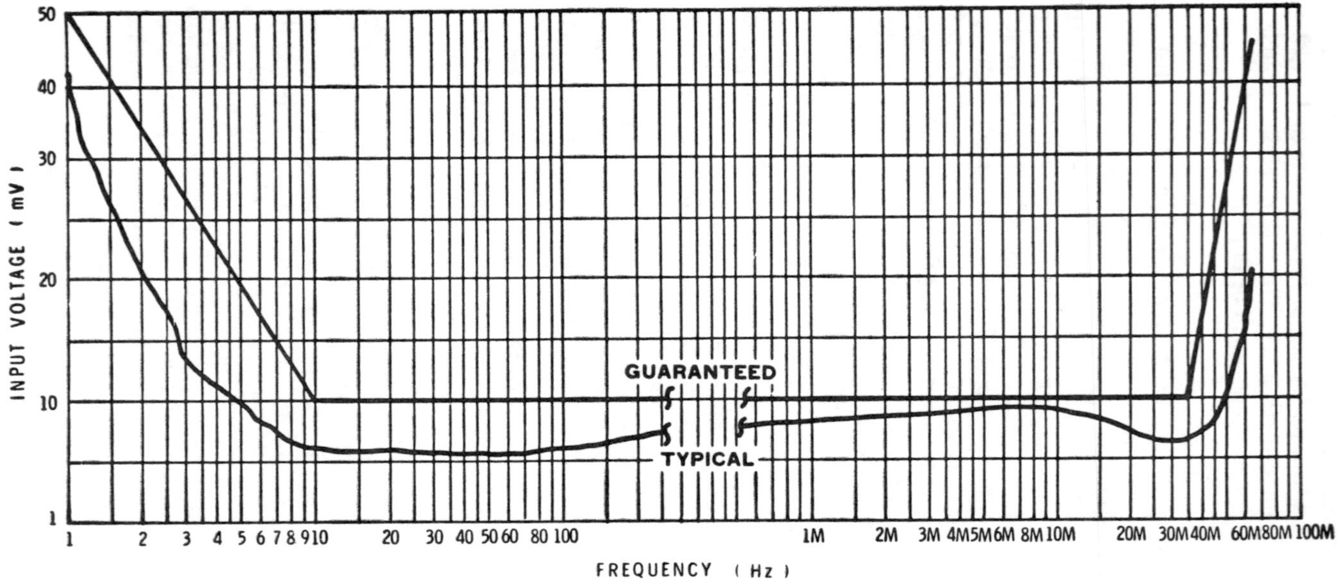

Fig. 3-5. The input sensitivity curve for the Heath SM-110 Series frequency counters. Note both guaranteed and typical curves are given. As shown by this set of curves, the typical sensitivity of a counter may be substantially better than the guaranteed limit. It is wise to know both figures, especially if a problem relating to input sensitivity is suspected of creating error.

even be destroyed. Manufacturers generally give some maximum voltage which must not be exceeded. This specification is accompanied by a set of derating curves for each input, showing how the maximum input voltage varies with frequency. Such a set for the Heath SM-110 counter is shown in Fig. 3-6.

One note of caution: 50-ohm inputs on frequency meters are *not* dummy loads. A transmitter or other high-level signal source connected directly to a frequency meter's 50-ohm input will almost always destroy the input circuit. This is an expensive mistake, as such inputs are often not field repairable.

Time Base

The main element determining frequency counter accuracy is the absolute accuracy of the reference frequency at the time of measurement. This is specified as a combination of a number of influences:

Set Tolerance: How precisely the oscillator is set to the intended frequency, usually given in parts per million (PPM). Low-cost instruments are generally set to between 1 and 0.1 PPM.

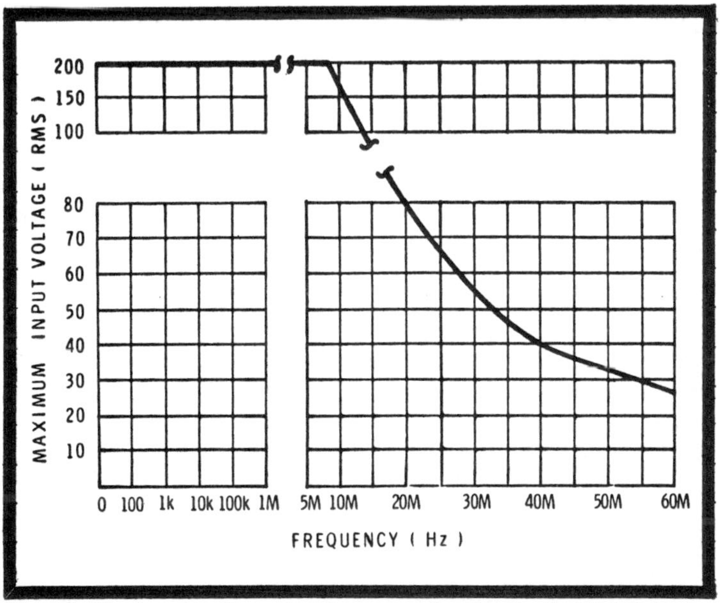

Fig. 3-6. The input voltage derating curve for the Heath SM-110 Series counters. Input levels in excess of the curve may destroy the input circuit components. Note that ample protection is given at 50–60 Hz against accidental contact with the 120V AC line.

Long-Term Stability: The stability in PPM per month (or year). Stabilities of 10 PPM per year to 1 PPM per year are typical for low-cost instruments.

Temperature Stability: The ability of the crystal to hold frequency over variations in temperature, given as PPM over temperature range or PPM per degree. Simple crystal oscillators will usually show variations of 5–10 PPM when ambient temperature varies from 10°C to 40°C (50°F to 104°F). Temperature-compensated crystal oscillators (TCXOs) and ovens can reduce such variations to 1 PPM or less. Beyond the 1 PPM level, oven-stabilized oscillators cost a significant portion of the frequency counter price.

Gating Intervals

Very often the available gating intervals that can be selected from the time base are specified as part of the instrument's capability. Most frequency counters will provide at least two gate intervals, one of them being 1 sec and the other being 1 to 10 msec, depending on the number of digits available. More gate intervals are often available and specified. A typical example is the 10 sec gating interval often provided to give 0.1 Hz resolution.

Display Time

The display time is the time between gating intervals. Usually this is no less than 200 msec and no more than 1 sec. Occasionally 10 sec gating intervals will produce a 10 sec display time; this situation leads to long waits to achieve high-resolution measurements. Some of the more sophisticated frequency counters permit the user to adjust the display time by a front-panel control.

FEATURES

Addiional features are the things that make an instrument convenient to use, readily adaptable to diferent situations, and expandable for future needs. Often such features cost little more but can shift a buy decision from one manufacturer to another. They include:

External-Oscillator Input. This input permits the user to replace a lower-stability internal oscillator with a higher-stability external oscillator. It also permits the use of such signals as 0.16666 MHz to derive measurements in the time unit of minutes instead of seconds. A typical example would be RPM.

Autoranging. Autoranging eliminates the necessity of changing the gating interval to obtain maximum resolution from the counter. The counter automatically ranges the gating interval to fill the complete display.

Optional High-Stability Clocks. Persons doing such critical work as making measurements for two-way radio service or making frequency certification measurements may desire a clock with high stability over time and temperature variations. Two types are generally available: (1) A TCXO (temperature-compensated crystal oscillator) uses a varicap–thermistor circuit to keep the crystal on frequency with temperature variations. The TCXO offers virtually zero warmup time and low cost for moderate stability. (2) The ovened oscillator is the best time base. Generally these are expensive; however, virtually any stability required can be purchased. Such oscillators may require anywhere from a few minutes to a few hours of warmup time.

Prescaling. Counters having prescaling, or the ability to add it, usually offer higher frequency range for lower cost.

BCD. Binary-coded decimal outputs permit the use of a printer or other remote display of measured frequency. Often combined with BCD output is remote programming. Remote programming permits selection of such items as inputs and the gating interval for some remote source. Both BCD and remote programming have advantages when used with automatic test systems.

Sensitivity and Level Controls. These permit adjusting the trigger point on the input signal to eliminate extra counts caused by unwanted noise. Figure 3-7 illustrates this. Note that the setting given in A would produce a frequency measurement double that of B.

Multiple Inputs. Multiple inputs permit making frequency ratio measurements, or provide a choice of input impedance.

Number of Digits. The number of digits indicates the highest frequency that can be displayed while maintaining full resolution. If a wide range of frequencies is to be measured regularly, a large number of digits means the gating interval will have to be changed less frequently to achieve the desired resolution. Some manufacturers offer the same instrument with two or three different numbers of digits. This feature should be carefully considered in light of future needs when purchasing an instrument.

Attenuator. If the input amplifier can easily be overdriven by high-level signals, an input attenuator is added. This is usually of the ×1, ×10, ×100 type.

Frequency Multiplier. This option permits the user to make high-resolution, low-frequency measurements in a short time. For example, a 1 sec gating interval produces a

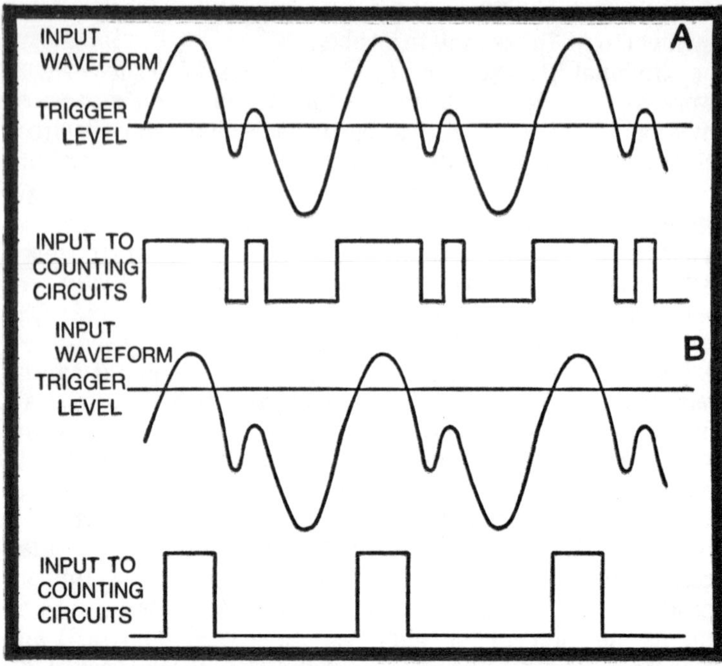

Fig. 3-7. The effect of trigger level on the measured frequency. (A) The input waveform causes two counts for each of the basic cycles of the measured waveform. (B) This is the same waveform with the trigger level or sensitivity level adjusted to give one counter per basic cycle.Whenever possible, it is best to observe waveforms on an oscilloscope prior to applying them to the input of a digital frequency counter.

resolution on a digital frequency meter of 1 Hz providing there is no prescaling. However, if the input frequency is first multiplied by 1000 and then measured, the resoluion will be 1 mHz (1/1000 Hz).

APPLICATIONS

The following is a sampling of frequency counter applications. This selection is not intended to be exhaustive, but it should well illustrate the specifications and operating principles of counters.

Transmitter Frequency Measurements

One of many applications is to determine the exact frequency of a transmitter. The first rule when using a frequency counter for this is to make the measurement without modulation. Modulation of any transmitter produces sidebands, and sidebands contain frequencies other than the carrier, which cause errors.

One popular method for coupling a signal into the frequency meter is to connect the frequency-meter input to a small whip antenna. In the VHF and UHF regions this may be a quarter-wave mounted directly on the BNC connector. However, any random length will work reasonably well at most high frequencies.

To measure a transmitter frequency, the frequency meter, with this coupling antenna attached, is moved to a spot near the transmitter where radiation is present, and the transmitter is activated. The proper distance (usually within a yard of the source) is that which will just produce a stable reading on the frequency meter. Closer coupling may produce too great a signal at the frequency counter input and result in erroneous readings from overloading or even damage to the instrument. Take care that the transmitter antenna does not come in direct contact with the frequency counter input.

Figure 3-8 shows a Heath Model SM-110-B VHF frequency meter being used to measure the transmitter frequency of a Heath HW-202-meter FM transceiver. If this technique is to be employed frequently, a whip mounted on a BNC connector should be obtained. Whatever you do, do not stick a wire directly in the input BNC of the frequency meter, as this will ruin the connector. If you do not have the combination BNC whip, a short length of coax with a BNC connector on one end and the braid stripped back a foot or so on the other end will do.

Note that the reading on the frequency meter indicates that the transmitter is 200 Hz above the intended frequency of 146.940000 MHz. Further analysis will tell us that 200 Hz is 0.0013% above 146.94 MHz. This is nearly five times closer tolerance than the 0.0005% required by the FCC in the neighboring 152–174 MHz commercial service. A common mistake made by many persons first using digital equipment is to call for more accuracy than is required, just because they have a digital meter.

Another handy device for use around transmitters and other circuits is a short piece of coaxial cable terminated by a sniffer coil. (See Fig. 3-9.) Three or four turns of insulated wire, such as No. 14, wound in a 2 in. diameter coil and connected across the end of a coaxial cable will suffice. The sniffer coil can then be coupled to the antenna, output tank circuit, final cage, dummy load, or some other point where some RF may be present. Again, the proper technique is to turn on the transmitter and bring the coil close enough to secure a stable reading.

Figure 3-10 shows a sniffer coil being used to measure the operating frequency of a dip meter. Note the end coupling that

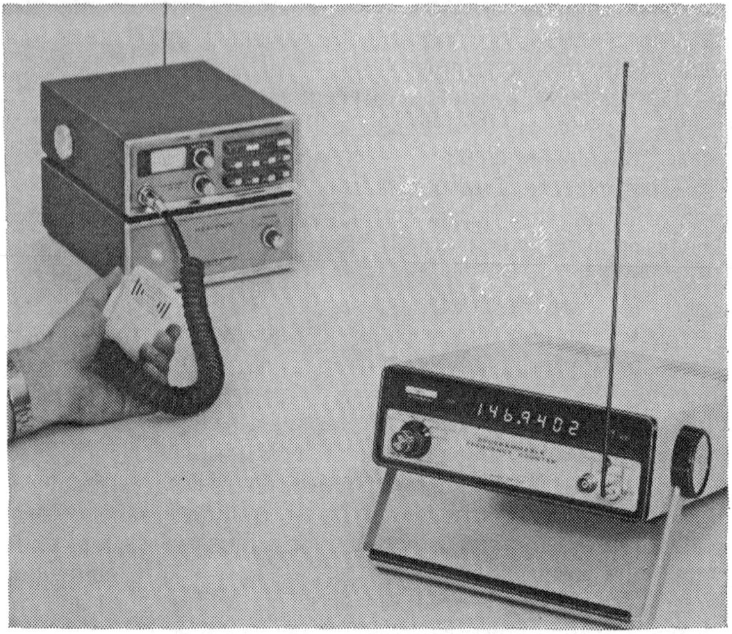

Fig. 3-8. A Heath–Schlumberger SM-110C 600 MHz UHF frequency counter is used to measure the transmitting frequency of the Heath HW-202 2-meter FM transceiver. A counter with sufficient sensitivity (50 mV or so) may be used with a whip antenna at some distance from the transmitting antenna. The correct distance is the greatest distance that will produce a stable reading on the frequency meter, and no less. The user must take care that the two whip antennas do not come in contact, or damage to the counter could occur.

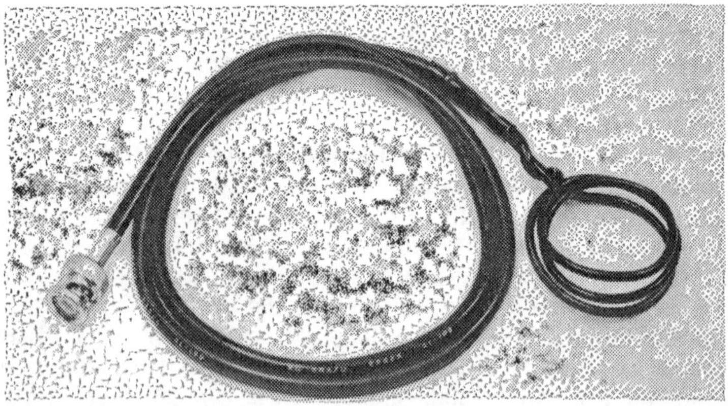

Fig. 3-9. The sniffer coil is an invaluable tool for extracting energy from circuits that cannot tolerate loading placed on them by a probe or cable. A few turns of stiff wire terminating the end of a coaxial cable will create this handy tool.

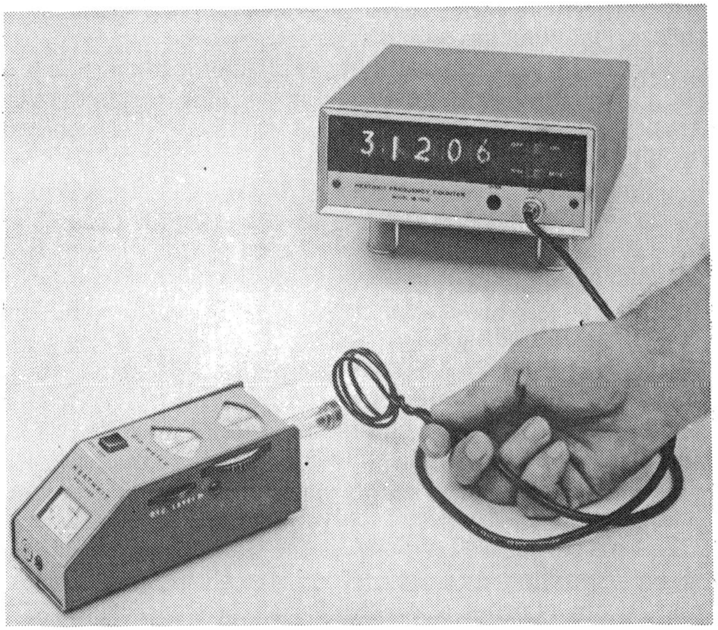

Fig. 3-10. The new Heath dip meter (HD-1250) demonstrates the ease of using the sniffer coil with the Heath IB-1100 30 MHz frequency counter. Loose coupling is the secret to good measurements, but sufficient input amplitude is required for a stable frequency reading.

is being used in this particular case. The secret to successful measurement with the sniffer coil is to get the reading with as little coupling as possible. Sniffer-coil techniques are particularly good around transmitters, oscillator tank circuits, crystals in crystal oscillators, and TV deflection circuits.

If no sniffer coil is available, one can be improvised by connecting the shield and center leads of a BNC to a clip-type test cable. Be careful not to allow the cable clips to contact hot circuit parts, or extra repair work may be necessary.

In measuring the output frequency of a single-sideband transmitter, a problem arises. Without modulation there is no output. *With* modulation the output consists of a band of frequencies. Even with single-tone modulation, the transmitted frequency is not the carrier frequency of the transmitter. Many single-sideband transmitters generate a CW signal that is offset by 1 kHz or so from the actual carrier frequency. To measure the output of a single-sideband transmitter accurately, the best technique is to temporarily unbalance the balanced modulator. Once this has been done, the previously mentioned techniques will produce a measurement of the exact carrier frequency.

Note: In some transmitters this is automatically done in the TUNE position. However, many sideband transmitters are not designed for continuous service, so measurement intervals should be kept short.

Another acceptable but more tedious technique is to make an in-circuit measurement of the frequency of each of the oscillators (fixed and variable) used in the heterodyning technique. Applying appropriate addition and subtraction to these frequency measurements will yield the exact frequency of the transmitted signal.

If the transmitter output frequency is greater than the capability of the frequency meter, there are a number of techniques that may be employed. With a transmitter of the multiplier type, such as many FM transmitters, there is a possibility of making the frequency measurement at a lower frequency stage and then applying the correct multiplier to this. This works well so long as the transmitter circuits are actually multiplying correctly. It is not a safe technique if there is any question of whether or not the transmitter's basic tuning procedure has been followed. Any frequency errors will be multiplied along with the signal. Percentage errors will of course remain constant.

A second technique that may be used is that of the *transfer oscillator*. Using a signal generator that is within the frequency range of the frequency meter, a harmonic from the signal generator is tuned into a receiver along with the unknown transmitter signal. The beat or difference frequency is reduced as close to zero as possible by adjusting the generator frequency. The frequency of the signal generator is then measured with the frequency meter.

Note: It is advisable to have the generator output level as high as possible in order to produce a high level of harmonics.

The measured frequency from the signal generator is then multiplied to arrive at the actual transmitted signal. Obviously some idea of the transmitter frequency is required to perform this measurement. For example, if a 27 MHz transmitter is to be measured and the frequency counter has a 15 MHz upper limit, the signal generator would be set at 13.5 MHz and the answer multiplied by 2.

Given a sufficient warmup time for stability, a grid dip meter may serve the purpose of the transfer oscillator. A sniffer coil, when placed over the grid dip coil, will couple sufficient energy into a frequency counter to produce a stable reading. The sniffer coil must be in position during the zero-beat process, as any movement of the sniffer on the grid meter will change the grid dip meter frequency.

The use of a transfer oscillator is a very powerful technique that will allow one to extend the frequency range of his frequency counter manyfold. There are certain cautions however: The user must be able to ascertain the exact multiple of the generator frequency that is in use; the generator must be sufficiently stable to remain within the receiver's frequency bandpass during the measurement; and the generator must be free of excessive FM, which would cause erroneous readings.

If a generator is used that has a tendency to drift slowly, the measurements still can be made. For example, suppose that a generator tends to drift lower in frequency. Set the generator at a higher frequency, and then, as the generator drifts across the frequency to be measured, watch for zero beat (or zero output from a discriminator on an FM receiver) and make the measurement. This method is often less tedious than trying to set a generator exactly on a particular frequency, especially if the generator is an older model and does not have the mechanical tuning resolution needed.

These techniques may be used for transmitter frequency measurements in the two-way FM, CB, ham, marine, and other services which require only that the technician be able to verify the accuracy of the measurement. Transmitter frequency measurements for the AM broadcast, FM, and TV services formerly required the use of FCC approved equipment, but this requirement has recently been removed.

Received-Signal Frequency Measurements

Although the signal level of a received signal is generally never sufficient to directly actuate a digital frequency meter, it is often desirable to determine the frequency of such a signal. One common technique has been discussed under *Transmitter Frequency Measurements*, that is, the use of a transfer oscillator. It is necessary to reduce the signal level from the transfer oscillator so that the signal generator does not block the receiver. The user must hear both the transfer oscillator (signal generator) and the distant signal. It may be necessary to remove the frequency meter and transfer oscillator to some location distant from the receiver so the signal generator may be turned to a high enough signal level to drive the frequency counter and yet not block the receiver.

An alternative technique is to measure the frequency of all oscillators within the receiver and any audio beat frequency if the signal is not zero-beat. The appropriate arithmetic manipulations are performed on these measurements, and the result is a number that is the received signal frequency. A device that automatically performs this function is a special-purpose frequency counter, the Health SB-650. This instrument automatically measures each of the oscillators in a Health SB-300 Series receiver and, by using up–down counters, it automatically subtracts and adds these frequencies to determine the actual received frequency.

An example of such a measurement, using the Health SB-303 receiver, is shown in block diagram form in Fig. 3-11. This indicates that when the receiver is tuned to (zero-beat at) 21.000000 MHz, the HFO (heterodyne-frequency oscillator) should be at 29.895000 MHz, the LMO (linear master oscillator) at 5.501400 MHz, and the BFO (beat-frequency oscillator) at 3.393600 MHz. Subtracting we have the following for an ideal receiver:

$$
\begin{array}{rl}
29.895000 & \text{HFO} \\
\underline{-5.501400} & \text{LMO} \\
24.393600 & \\
\underline{-3.393600} & \text{BFO} \\
21.000000 &
\end{array}
$$

However, if the oscillators were measured exactly with a frequency counter and found to be as follows,

$$
\begin{array}{rl}
29.894500 & \text{HFO} \\
\underline{-5.5000000} & \text{LMO} \\
24.394500 & \\
\underline{-3.393400} & \text{BFO} \\
21.001100 &
\end{array}
$$

the actual received frequency would be 21.001100 MHz, or 1.1 kHz high.

Another technique often used with high-stability single-sideband receivers is to assume the exact frequency of all crystal oscillators and only measure the VFO frequency. From this frequency measurement and a knowledge of bandswitch settings, the frequency setting of the receiver is determined. This technique can lead to some error, as crystal oscillators are not always exactly on frequency. A method for partially overcoming this difficulty is to develop a history of the frequency drift of each of the crystal oscillators in the receiver. After aging, the crystals will become quite predictable, and at a constant ambient temperature, the future value of their frequency can be calculated quite closely.

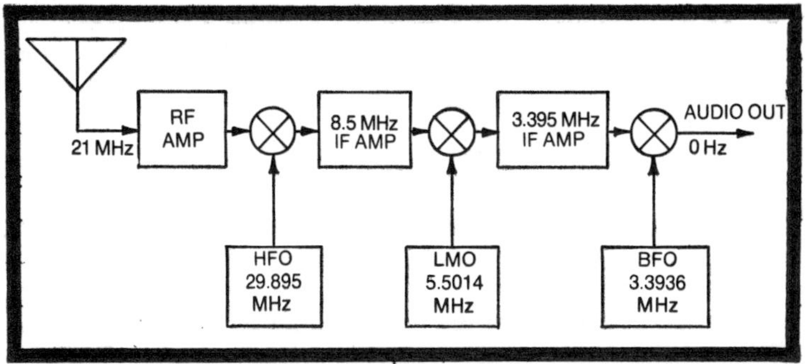

Fig. 3-11. A block diagram of the Heath SB-300 amateur radio receiver, showing the frequencies used to receive a frequency of 21 MHz.

A note of caution should be brought out here. The digital frequency counter tends to be a somewhat noisy instrument. Starting with a 1 or 10 MHz crystal oscillator and following with numerous decades of digital division, the counter generates harmonics that cover a large portion of the RF spectrum. Although attempts are made by manufacturers to reduce the noise radiation to the lowest possible levels, it is often not practical to shield the instrument to the extent of preventing interference with receivers having 0.1 μV sensitivity.It is therefore advisable, when using a frequency counter near a high-sensitivity receiver, to remember that the frequency counter will often generate birdies and spurious responses within the receiver. When using the receiver at its highest sensitivity, the frequency counter should be shut down.

In-Circuit Measurements

To the experimenter the frequency counter is a valuable tool for analyzing experimental circuits. Each experimenter will build up his own techniques for making frequency measurements in experimental circuits. There are, however, a few standard techniques worth noting.

When working in RF circuits around crystals and coils, the sniffer probe mentioned in earlier sections is extremely valuable. Generally this is connected to the 1M input; however, when the results are not as desired, the 50-ohm input should be tried (if the frequency counter has one). When using the sniffer probe, couple as lightly as possible to the coils. The best technique is always to avoid loading the circuit under test any more than possible. To couple to a coil, either bring the two coils parallel to each other or couple near the end of the coil. Sometimes, by moving the sniffer coil around the circuit, a point may be found near a resistor lead or a capacitor where there is sufficient radiated energy to make a measurement.

When trying to measure the frequency of crystal oscillators, it is always advisable to measure at least one stage away from the crystal to avoid "pulling" the crystal off its frequency. If this is not possible, use the sniffer coil by lowering it over the crystal or, again, try to couple it to various portions of the circuit.

Another extremely handy device for use with counters that have a 1M, low-capacitance input is an oscilloscope probe. Such a probe increases the input impedance by a factor of 10, to 10M, and reduces the capacitive loading to the 3 pF area. This probe permits direct coupling to many higher impedance circuits which would not tolerate direct coupling with clip leads.

Although loading of the circuit is decreased by a factor of 10, there is a disadvantage. The sensitivity of the frequency meter is also decreased by a factor of 10. If the frequency counter is to be used frequently with a ×10 probe, a high-sensitivity counter should be purchased. A decrease in sensitivity is not always a detriment, however. Occasionally the measurement of high-level circuits will be required, and such a probe will extend the voltage range of the input circuits by a factor of 10. Such an oscilloscope probe also serves another purpose. The probe has a clip mechanism that is simply connected onto circuit components, and this often saves the extra work of replacing components damaged by shorts caused by more clumsy methods of connection.

One of the most common cables for making in-circuit measurements is the coaxial cable terminated with alligator clips. For many circuit applications this is entirely satisfactory. Remember, however, that coaxial cable has a capacitance in the vicinity of 30 pF per foot, so a 3 ft cable will load the circuit by approximately another 100 pF when connected to a frequency counter. While in some circuits such additional loading may not cause any harm, in others it is disastrous.

A technique for reducing this capacitance is to couple the hot lead of the frequency counter by means of a small capacitor between the alligator clip and the circuit point. There are occasions when making high-frequency measurements where capacitors as small as 1 pF are usable. Here one must remember that the signal is being applied to a capacitive divider, and the resultant loss in sensitivity can be great. However, if sufficient signal is available to use this technique, the circuit loading is negligible.

When making direct connections, one should also be careful of DC voltages present at the same point in the circuit. Often AC-coupled frequency counters use input capacitors with 50–100V DC breakdown ratings. Some tube circuits, such as found in transmitters, contain sufficient voltage to destroy these capacitors. Again, the oscilloscope probe will extend the DC capability by a factor of 10.

When making frequency measurements from logic circuits, it is wise to either use an oscilloscope probe or to use a series capacitor with the coaxial cable. Many logic circuits, such as TTL flip-flops, will not tolerate 100 pF capacitive loads on their outputs; it often causes circuit malfunctions. When measuring the frequency of signals in logic circuits and using a frequency counter which has both AC and DC input coupling, be sure that the counter is in the AC position. Logic signals generally do not swing through zero, but operate between a certain DC voltage and some more positive or negative DC value. Such signals will not trigger a DC-coupled frequency counter, as it depends on the signal passing through 0V.

A signal commonly found in logic circuits is a narrow spike at a low repetition rate. Remember that the sensitivity of a digital frequency meter is specified as an RMS value. For example: A 2.5V, 100 nsec pulse repeated at a 1 msec rate will yield an RMS value that may be below the sensitivity level of the frequency counter.

SOURCES OF ERROR

As wih any instrument, it is possible to make measurement errors with a digital frequency counter. Sometimes these errors are subtle, such as a minor loading of a crystal oscillator. Other times they are obvious, such as overdriving the input and causing the frequency of a transmitter to appear double its actual value. Errors of frequency measurement are caused by the inaccuracies of the frequency meter itself or by improper measurement techniques.

Time base

The time base of the frequency counter creates the major error that the counter itself contributes to the reading. As an example, if the time base frequency is low by 10 PPM, then the 1 sec gating interval will be extended by 10 μsec. A measured signal will actually be measured for 1 sec plus 10 μsec, and the frequency reading will be high by 10 PPM. The basic accuracy specification of a frequency counter is usually given as plus or minus one count (the phasing ambiguity), and plus or minus the time base error; the time base error usually being the larger portion.

It is well to remember that a counter very frequently can have far more resolution than the time base error will permit you to use. For example, suppose a 450 MHz FM transmitter's frequency is being measured by a frequency counter with an accuracy of 1 PPM. This means that the measured frequency is going to be in error by ±450 Hz. However, the frequency counter may easily be able to resolve 10 Hz or even 1 Hz—that's 450 times greater resolution than absolute accuracy!

Phasing Ambiguity

Another error always present in a digital frequency counter is the ±1-count error, or phasing ambiguity. As a digital frequency counter cannot count fractions of a cycle, it must pick one point in the cycle at which it notes the presence of a cycle. One can see from Fig. 1-12 that it is possible, depending on the phase relationship between the gating signal and the unknown frequency, for the number of counts to vary by one count from gating interval to gating interval. As the user does not know whether the gating interval was just under or just over the width required to produce an additional count, this leads to a ±1-count ambiguity, which must be added to the time base error. This form of error has the greatest significance at low frequencies. For example, if a 10 MHz signal is measured on a 7-digit frequency counter for 1 sec, the displayed frequency might be 9.999998 MHz. The ±1-count ambiguity means that this measurement actually could be 9.99999 or 9.999999. However, if this frequency meter had a ±1 PPM time base error, the actual frequency could be anywhere from 9.999989 to 10.000008 MHz, and the phasing error is not important. If, however, a measurement is made of a 100 MHz oscillator and the 10 sec gating interval of a frequency counter is used, the readings might vary from 99.9 to 100.1 or ±0.1 MHz due to phasing error. However, the 1 PPM error of the time base would only contribute a 0.0001 MHz error to the readings. As can be seen by this example, phasing ambiguity can, at low frequencies, cause a high percentage of error.

Errors of Averaging

A more subtle problem is the error of averaging. If during a 1 sec gating interval the frequency of the measured signal varies, the frequency counter will display the exact number of cycles within 1 sec, which is the average frequency during that time. Such an error may be detected by changing the gating interval. If the displayed frequency appears to vary when going from a 1 sec gating interval to a 1 msec gating interval, for example, one should suspect some frequency modulation of

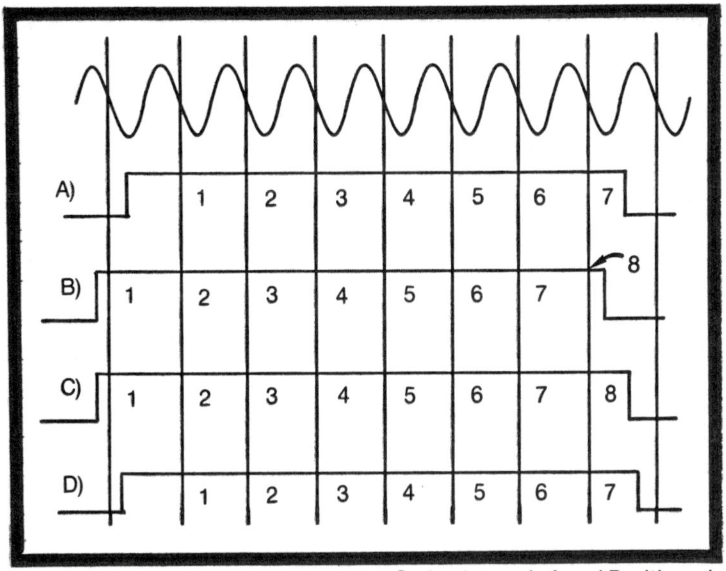

Fig. 3-12. Effects of phasing ambiguity. Gating intervals A and B, although of the same length, can produce seven or eight counts, depending on the phase relationship between gating interval and input signal. Gating intervals C and D produce eight and seven counts, respectively; but if they were slightly longer, they would produce nine and eight. The user does not know if A, B, C, or D represents his particular case, and therefore he must assume eight counts, plus or minus one count.

the measured signal. A signal that appears to be unstable with a short gating interval but stable with a longer gating interval is most likely being averaged by the longer gating interval.

Improper Signal Level

Improper use of equipment can cause many measurement errors. One of the most common misuses is failure to provide the proper signal level. Insufficient signal level may cause the frequency counter to register on noise and, at other times, to miss the signal completely. In overdriving the input of the frequency meter, the circuits under test or the input amplifier of the frequency meter may double or otherwise multiply the input frequency artificially. Both conditions can easily be tested for by varying the input level. A good rule to follow is never use a greater signal than the least signal required for a stable reading.

Unterminated Cables

The use of unterminated coaxial cable about the length of an electrical quarter-wave at the frequency being measured can result in a counter de-densitization. For example, a 3 ft length of RG-58 connected to the 1M input of a frequency counter will result in a virtual short circuit at the cable input at 54 MHz. This short circuit will severely attenuate the signal being measured and may cause errors identical to those caused by insufficient signal level. Whenever possible, coaxial cables should be terminated in their characteristic impedance. Coaxial cables of 50 ohms may be terminated by using the 50-ohm input of the counter or an in-line coaxial cable terminator connected to the 1M input of the counter. If these terminations are not available, a 47-ohm resistor shunted across the cable will suffice. Often it is undesirable to have the input impedance as low as 50 ohms, and therefore the coaxial cable is connected to the 1M input and used as a shielded lead. When using the input cable in this way, remember that a problem may be caused when the length nears an electrical quarter-wavelength. If the 1M input is required and the test cable being used is close to an electrical quarter-wavelength, the cable should be either lengthened or shortened to avoid this situation.

Transmitters and Errors

High-powered transmitters, especially those improperly grounded, may cause errors in digital frequency counters. Stray radio-frequency interference (RFI) that is sufficiently intense in the vicinity of the frequency counter may, by direct radiation as well as conduction through the power lines, cause improper operation of the counters, control circuits, or time base. Such a condition may cause a complete malfunctioning of the frequency counter or improper readings. Interfering signals of less intensity superimposed on the signal being measured will cause erroneous counting. Errors may occur even if the interfering signal is beyond the frequency capability of the instrument. Radio-frequency interference should be suspected if the frequency counter, with no input cable, displays readings that are different when the transmitter is turned off than when it is turned on. Remedies for such interference include shielding and bypassing of parts associated with the frequency counter and perhaps similar corrective measures for the transmitter itself.

Ground Currents

Unwanted currents of sufficient amplitude that are superimposed on the signal being meaured will always cause some form of measurement error. A common form of this error is caused by 60 Hz currents generated between instruments that have improper grounding. In such instruments a circulating current may develop because of a potential difference between the grounds of the two chassis. This current will circulate in a loop through the braid of the coaxial cable and the grounding system of the instruments. Currents of 1–10 mA can easily cause 60 Hz interference signals of 4–15 mV. Frequently these signal levels may be of the same order of magnitude as the signal being measured, and the 60 Hz interference will in some manner modulate the signal being measured and cause errors. This condition may be detected by changing the gating interval or changing the grounding system of the various instruments.

Noise

Errors in measurement occur when frequency measurements are made from sources that have noise at their transition periods. A typical example of this is contact bounce. With the exception of mercury-wetted contacts, contact closures will result in a short burst of high-frequency noise at the time of contact opening and closing. An *RC* filter may be used to reduce the level of these signals below the sensitivity of the digital frequency counter. When using logic circuits, which also tend to respond to contact bounce, it is often best to use a set of cross-coupled gates or a monostable multivibrator for contact-bounce elimination. If a logic circuit incorporates contact-bounce prevention, frequency measurements should be taken from the "clean" side of the contact bounce eliminator.

Sensitivity

A frequent point of confusion concerns the sensitivity of the digital frequency counter with respect to the output amplitudes of RF signal generators. Radio frequency signal generators are usually intended to produce levels from the submicrovolt level to the 100 mV or perhaps 1V level. The digital frequency counter will only be able to measure signals from the highest one or two decades of output level from the signal generator. To overcome this problem, some generators have an external output that is not attenuated. This output may be monitored continuously by a digital frequency counter to give the exact frequency of the generator. Once this output is connected to a cable, the effectiveness of the shielding built into the generator is lost, and radiated signals far in excess of the attenuated output may exist.

Excessive Resolution

Extended resolution may be made when making high-frequency measurements with a digital frequency counter with a limited number of digits by selecting a number of gating intervals during the measurement. For example, to measure the output frequency of a 450 MHz FM transmitter on a 6-digit frequency meter, the following procedure would be used: The first reading would be taken using the 1 msec gating interval, and the displayed reading might be 449.452 MHz. Changing the frequency counter to the 1 sec gating interval might result in a reading of 451,864 kHz. Combining these two readings and realizing that the digit 2 of the first reading resulted from an upward rounding off of the digits 1-8 of the second reading, the transmitter frequency is read as 449.451864 MHz.

Added resolution obtained by changing the gating interval is valid only to a point. If a frequency counter with a time base known to be within 0.1 PPM gives an absolute accuracy of measurement of the 450 MHz signal in the example above that is plus or minus 45 Hz, the last 64 Hz is in serious question. However, if a second transmitter is measured and the frequency is found to be 449.451800, a determination that the frequency difference between two transmitters is 64 Hz is quite valid. The resolution provided by such a measurement exceeds the absolute accuracy. However, if two measurements are taken within the timespan of the short-term stability specification of the oscillator, frequency difference measurements based on this extended resolution are quite accurate. Oven oscillator and TXCO specifications very frequently give short-term stability specifications, indicating the drift of the crystal over a few seconds. The short-term stability can be used to determine the validity of high-resolution measurements.

CALIBRATING THE FREQUENCY COUNTER

The long-term stability of the crystal oscillator with aging and the user's need for accuracy are the prime considerations in determining calibration intervals for the digital frequency counter. A frequency counter with a time base that is kept within a few parts per million will suffice for all but the most exacting needs.

It is well to remember that, as measurement devices go, the digital frequency counter has exceptional accuracy. It is quite easy to achieve measurements with an accuracy of 1 PPM. This corresponds to 0.0001%, which in most fields is an extremely good limit of error. Too-frequent calibration is a common error made by many frequency counter users. When calibration does become necessary, it may be done in a number of ways.

Radio station WWV can generally be received in all parts of the United States at either 5, 10, or 15 MHz. In the following technique the frequency counter is placed close enough to the receiver that the time base oscillator creates a beat note with the received WWV signal during periods of no modulation. It is preferable to use the highest frequency of WWV that can be received, as this will permit the highest degree of accuracy for the calibration. During the period of no modulation, the FREQUENCY control for the oscillator is adjusted for zero beat with WWV. WWV is generally received with a frequency accuracy of 0.01 PPM as long as there is not a line of dusk or dawn between the receiving location and WWV. More precise signals are available from WWVB and WVL at 20 and 60 kHz. Specialized receivers are required for these two stations, and in all but the most sophisticated of applications, their accuracy is not required.

If WWV is not available, a 10 kHz subharmonic of the time base oscillator may be used to create a beat note with an AM radio station. These radio stations usually maintain frequency accuracies of the order of 1 PPM and therefore provide a good reference. Input sensitivity and frequency range limits of a frequency meter can be checked using laboratory RF signal generators. Front-end sensitivity usually will not have to be readjusted; and if it is required, the technique varies from instrument to instrument and is usually described in the manufacturer's service manual.

SUMMARY

The frequency counters discussed here have been confined to the low-cost units that are making their way into new places in industry and into the hands of experimenters. Many of the characteristics that have been discussed are equally valid for the very expensive, sophisticated units found in research laboratories. Remember, however, that laboratory units will present some special characteristics that are unique to their price class and have not been covered here.

The future should tend to bring even lower cost frequency counters into being and certainly will bring more sophistication at the present prices. The fundamental concepts and measurement techniques will remain the same; however, the user will have to pay constant attention to keep abreast of the new equipment.

Chapter 4 Oscilloscope

The oscilloscope is often labeled the most valuable tool in the field of electronic measurement. Still, many people use an oscilloscope every day without understanding its basic principles or specifications. Consequently they fail to operate this instrument effectively.

Perhaps one of the reasons the modern technician falls short in his knowledge of oscilloscope operation is that the price of oscilloscopes, while increasing slightly at the very bottom of the line, has been coming down dramatically for high-performance instruments. The result of this change has been the virtual elimination of the sub-$100 oscilloscope and the prevalence of the high-performance oscilloscope in the under-$1,000 classification. Yesterday's television service repairman either made do without an oscilloscope, or more likely, made do with a 3–5 MHz oscilloscope with recurrent sweep. This oscilloscope was probably purchased for between $50 and $250. Today, for slightly over $200, the service technician can purchase a 10 MHz triggered oscilloscope incorporating a calibrated vertical attenuator and time base. Suddenly the technician is face to face with sweep magnifiers, automatic and normal triggering, trigger level control, and a host of other specialized functions which did not exist in the recurrent-sweep oscilloscope.

Why make such a necessary tool so complicated? The reason is that an increased complexity of controls increases the scope's already great versatility.

This chapter reviews the fundamentals of an oscilloscope, examines specifications and features in detail, reviews operation of basic oscilloscope sections, and develops a series of applications for oscilloscopes which exemplify the various modes of operation.

It is not within the scope of this book to discuss all the various types of oscilloscopes available today. Discussion is confined to instruments in the sub-$1,000 class, mostly service instruments. Such items as specialized plug-ins, spectrum analyzers, and time domain reflectometers are not discussed.

FUNDAMENTALS

The old saw "A picture is worth 10,000 words" is directly applicable to the oscilloscope. The oscilloscope presents a graphical display which, in most cases, shows how a voltage changes with respect to time. For the simplest oscilloscope, not only must this voltage change in time, but it must also repeat these changes regularly, or *periodically*. However, the vast majority of signals do vary periodically.

Of course a graph can be made by taking the voltage measurements with some other instrument, say a digital voltmeter, and then plotting these numbers. A reading can be taken at 1:00 o'clock, 1:10, 1:15, and 1:20, and these can be plotted on a graph as shown in Fig. 4-1. This technique is tedious and is not easily applicable to rapidly changing signals. The oscilloscope provides a graph with much greater speed and much less tedium.

Figure 4-2 is a block diagram of a basic oscilloscope. The oscilloscope consists of a cathode-ray tube (CRT), a vertical amplifier, a horizontal amplifier, and sweep circuits.

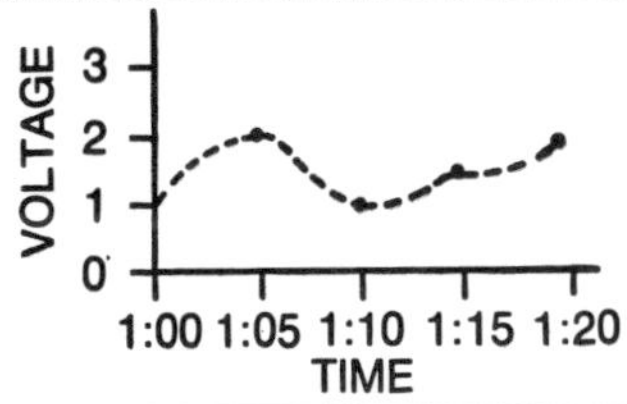

TIME	VOLTAGE
1:00	1.0
1:05	2.0
1:10	1.0
1:15	1.5
1:20	1.8

Fig. 4-1. A hand-plotted curve. As there are only five data points, the smooth curve must be assumed and sketched in by hand. The oscilloscope gives an electronic mechanization of this process.

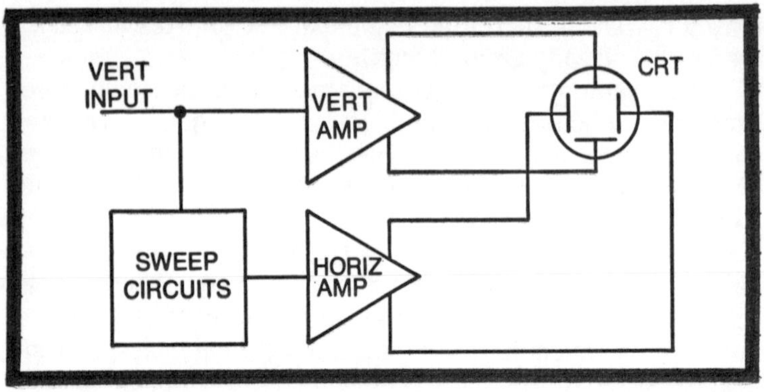

Fig. 4-2. A simplified block diagram of an oscilloscope.

Amplifiers

Vertical and horizontal amplifiers increase small electrical signals sufficiently to deflect the electron beam vertically and horizontally over the CRT screen. Typically a CRT will require a voltage between its vertical plates of 100V to deflect the beam across the tube. Oscilloscope input levels will be more like 100 mV, thus requiring a gain of 1000 in the amplifier.

Sweep Circuits

A sweep circuit is used to cause a horizontal displacement of the beam on the face of the CRT that is proportional to time. The sweep circuit generates a waveform (Fig. 4-3) which rises linearly with respect to time and then suddenly drops to zero;

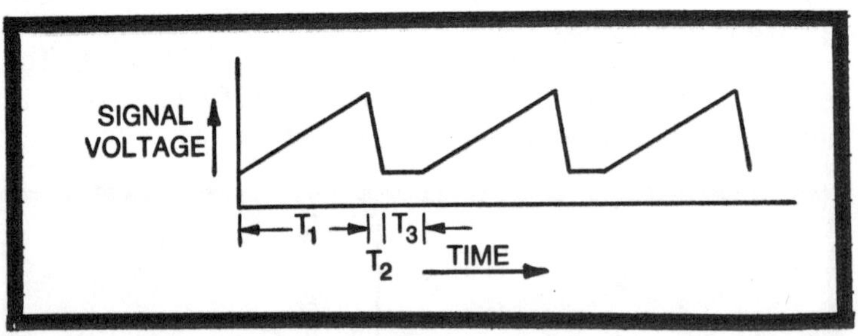

Fig. 4-3. Three cycles of the waveform generated by the sweep circuits of an oscilloscope. The duration indicated by t_1 is the highly linear ramp which deflects the beam across the face of the CRT, creating the time axis of a graph. The duration indicated by t_2 is the retrace time, when the beam is being moved back to its starting point. The duration indicated by t_3 is the holdoff time. Holdoff time is zero on a recurrent sweep oscilloscope and a fixed value dependent on sweep speed, plus a random variable time for triggered oscilloscopes.

it waits for some definite time, then begins to rise linearly again. This signal is commonly called a sawtooth or ramp.

Time Base

The ramp signal is used to drive the horizontal amplifier. The initial beam position will be at the left edge of the CRT. As the sweep circuit's output voltage increases, the beam is driven to the right, when the peak of the ramp is reached, the beam is at the right edge of the CRT. The beam then rapidly returns to the initial point as the sweep voltage returns to zero. When the horizontal amplifier is driven with such a linear sweep signal, the horizontal displacement of the beam may be equated to time. This gives the oscilloscope a time base. For example, the center of the CRT represents 0.5 sec of elapsed time with respect to the left edge of the CRT if the period of the ramp is 1 sec.

The ramp must be synchronized to the vertical signal. In other words, it must start at the same point every time on the input waveform. If this did not happen, the displayed waveform would start randomly, and a jumbled display such as that in Fig. 4-4 would result.

Recurrent Sweep

To prevent the condition shown in Fig. 4-4, synchronization, or triggering, circuits are employed. In the simplest oscilloscopes the sweep generator consists of a free-running oscillator which generates the ramp waveform. The free-running oscillator is synchronized to the vertical

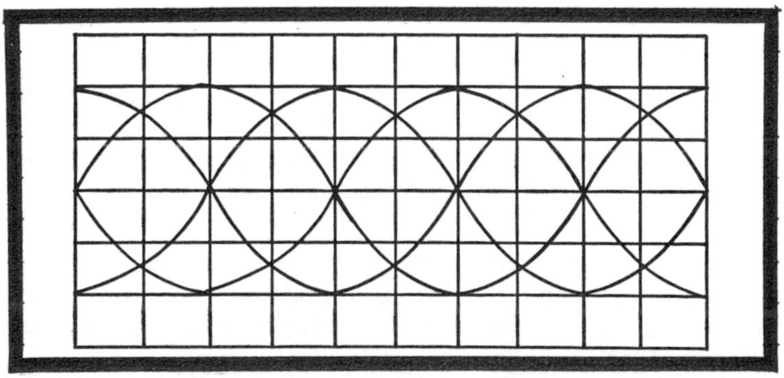

Fig. 4-4. An oscilloscope display of a sine wave without synchronization between the vertical amplifier and the horizontal sweep circuits. Although this diagram only shows four traces for illustrative purposes, an actual oscilloscope will essentially show a blur, as many more traces will start at random points.

signal by adjusting the sweep generator until its frequency is identical to, or a submultiple of, the frequency of the vertical signal. With the oscillator and vertical-amplifier signals at approximately the same frequency, a small amount of the vertical signal is injected into the sweep oscillator. The sweep generator will lock or synchronize, tending to assume the frequency and phase of the vertical signal. Obviously this method of synchronization is somewhat crude; however, until recently it has provided the lowest cost method for deriving a synchronized horizontal time base. For some forms of work this is entirely satisfactory. This elementary time base, called *recurrent sweep*, has some rather severe limitations: It can only synchronize an integral number of vertical cycles; it is virtually impossible to change the point at which the triggering starts; and unless some separate form of time measurement is available, the period of the sweep signal is unknown.

Triggered Sweep

All of these problems may be solved through the use of a calibrated triggered sweep. With triggered sweep, the sweep generator is converted from a free-running oscillator to a single-shot generator; that is, the oscillator will produce one single ramp of a known period when the appropriate trigger signal is applied. With this time base the user may select the exact period of sweep and, through various amplification and comparing circuits, the exact point on the input wave at which he desires the sweep to start. Given a time base of this sophistication, there are may applications for the oscilloscpe which were simply not available with recurrent sweep.

Blanking

The rapid retrace of the beam from the right side of the CRT to the left can cause undesirable traces on the display. To eliminate this effect a special signal is developed just before the end of each sweep signal. This signal *blanks*, or turns off, the beam in CRT. The blanking signal prevents the undesirable retrace signals from appearing on the face of the CRT. Normally the trace is blank until a new sweep is under way. This permits the ramp to become fully linear, which it may not be in the first few percent of its rise.

Some oscilloscopes have provisions for external input to the blanking circuits so that the blanking may be controlled externally for special applications.

Dual-Beam Oscilloscope

On the simple oscilloscope of Fig. 4-2, the vertical movement of the beam is governed by the signal in the vertical amplifier. Such an oscilloscope is called a single-trace oscilloscope. This simply means that all vertical deflection is a function of one input to the scope.

Frequently, when working with an electronic circuit, it is desirable to determine cause and effect—that is, to compare the input and output signals of the circuit. This may be done with a single-trace oscilloscope by first observing the input signal, then moving the probe to the output and observing the output signal. Obviously this process is difficult. As neither trace remains on the screen permanently, one has to remember each small variation of each waveform to compare them. One method of overcoming this is shown in Fig. 4-5.

This configuration is known as a dual-beam oscilloscope. It employs a special CRT which has two sets of vertical-deflection plates and two electron guns but only a single set of horizontal plates. Both electron beams are simultaneously deflected from left to right by the horizontal amplifier, and therefore, by the sweep signal. However, one of the beams will respond

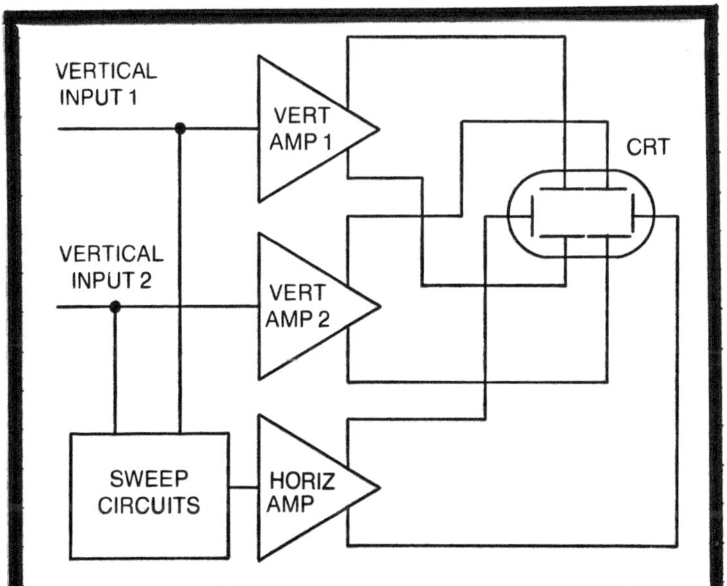

Fig. 4-5. A simplified block diagram of the dual-beam oscilloscope. Note that a specialized form of CRT is employed which contains two electron guns and two sets of vertical-deflection plates. Both beams are under the influence of one set of horizontal-deflection plates.

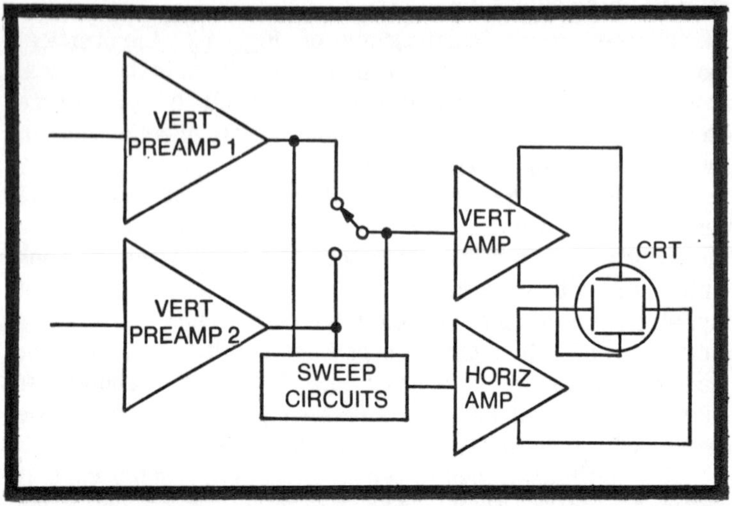

Fig. 4-6. A simplified block diagram of the dual-trace oscilloscope. Note that a standard CRT with a single set of horizontal- and vertical-deflection plates is used. Signals to operate the vertical switch are derived from an oscillator in the chopped mode from the sweep signals in the alternate mode.

vertically to the signal applied to vertical amplifier number 1 and the other will respond to the signal applied to vertical amplifier number 2. The dual-beam oscilloscope overcomes the problem of moving the probe from input to output. As could be expected, the dual-beam oscilloscope costs additional money, although a few are available below $1 000.

Dual-Trace Oscilloscope

This problem has been more popularly solved with the dual-trace oscilloscope (Fig. 4-6). This oscilloscope is similar to the single-trace oscilloscope; however, two vertical preamplifiers are employed, and a switch connects the main vertical amplifier to either input 1 or input 2. This switch is controlled by the time base circuitry or an internal oscillator. The switch operates in one of two modes.

In the *alternate mode* the switch is changed from one input to the other following each horizontal sweep. On the first sweep, input 1 is displayed; on the second sweep, input 2 is displayed. On the third sweep, input 1 is again displayed, and so on. The alternation is rapid enough to keep the trace presented by one input from fading during the time the other input is writing.

At sweep rates of sufficiently long duration to cause a flickering problem, a different mode of dual-trace display is used. In this, the *chopped mode*, the switch is no longer operated by successive cycles of the time base, but is alternated by a high-frequency oscillator, generally in the 50–200 kHz region. This produces a trace as demonstrated in Fig. 4-7. The time intervals in the figure are considerably exaggerated. In actuality, even at quite short time base periods, one must carefully scrutinize the trace to observe the chopping effect.

The dual-trace configuration is popular because it does not require a special-purpose, and therefore expensive, CRT. The dual-trace technique offers an additional advantage in that it may be expanded to more than two channels, whereas the dual-beam scope would require an additional set of deflection plates and an additional electron gun assembly in the CRT for any additional channel.

Trigger Selection

With either the dual-trace or the dual-beam oscilloscope a question presents itself: Where should the trigger signal be taken from? Most oscilloscopes using dual vertical channels have a switch permitting the user to select either input 1 or input 2. In some products the user is often offered a third option of triggering on the combined signal after the switch.

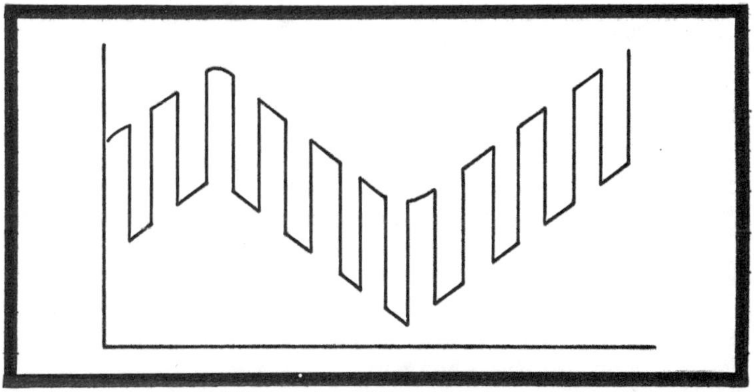

Fig. 4-7. An expanded illustration of the chopping technique for displaying two separate signals (a sine wave and a triangle wave) on a dual-trace oscilloscope. This technique is used when the writing rate using alternate modes will allow one trace to fade before the other has been completed. The chop rate is usually sufficiently high. There are at least 100 samples of each waveform per sweep. The vertical transistions between traces will either be blanked or made at such a high speed that the beam cannot write.

Electronic Switch

The simple single-trace oscilloscope may be converted to dual-trace operation by means of an external electronic switch. These electronic switches have not been too popular, as they tend to limit the portability of the oscilloscope, and the cost of single versus dual trace on a high-performance oscilloscope is generally low.

The electronic switch provides separate input attenuators for each channel to be observed, some amplification, and separate POSITION controls for each channel to permit vertical trace separation. Electronic switches are available in 2- and 4-channel configurations.

Delay Lines

If the triggered oscilloscope is used with very short periods of horizontal sweep (generally under 10 μsec or so) and the user desires to observe either the leading or trailing edge of a pulse in some detail, there is a new problem. Most trigger circuits are incapable of producing any sweep before the first few hundred nanoseconds of the vertical signal has actually arrived at the vertical-deflection plates. Therefore the signal that caused the triggering—say, the leading edge of a fast pulse—generally reaches the deflection plates a few hundred nanoseconds before the sweep actually starts. When this occurs the initial part of the signal is completely lost. To avoid this the signal in the vertical amplifier must be delayed. The most common technique is to use some form of transmission line to slow the signal. The signal in the vertical amplifier must be delayed enough to permit the trigger and then the sweep circuits to respond. To achieve these considerably long delays without limiting the frequency response of the vertical amplifier, rather specialized transmission lines are employed. Delay lines certainly add to the oscilloscope's performance and obviously add to its cost.

XY Oscilloscopes

Occasionally it is desirable to make certain measurements by comparing two external signals, one of them on the *Y* or vertical axis and the other on the *X* or horizontal axis. When the oscilloscope is used in this mode, the internal sweep circuits are no longer employed. With the simple oscilloscope, *X Y* operation is achieved by applying one signal to the vertical input and the other signal to the external-input connection of the horizontal amplifier. Horizontal amplifiers do not usually have calibrated attenuators, nor will they accept a wide range of input signal levels. When the oscilloscope is either a dual-trace or dual-beam type, two identical vertical amplifiers are available. Such oscilloscopes often have an *XY* mode, in which one of the vertical-input amplifiers remains connected to the vertical-deflection plates, and the other vertical-input amplifier is connected to the horizontal plates. This provides the user two wide-range inputs with calibrated attenuators.

Sampling Oscilloscopes

The oscilloscopes discussed above will generally suffice for most servicing and development work. However, there are certain types of measurements which require very high frequency oscilloscopes. The bandwidth of the vertical amplifier limits the ability of the oscilloscope to observe high-frequency signals. To overcome this problem, a specialized oscilloscope has been developed for viewing periodic waveforms of extremely high frequency. This technique is called sampling. The sampling oscilloscope takes a small sample of the high-frequency waveform each time a horizontal sweep occurs. After a number of horizontal sweeps, the point of sampling is moved slightly, and sampling occurs again. The result of this technique is an output from the sampling gate which is a replica of the high-frequency waveform but whose frequency is very much lower. Figure 4-8 shows an exaggerated display from a sampling oscilloscope. These oscilloscopes are very specialized instruments. They tend to be quite expensive and difficult to operate. However, when a waveform must be observed and its frequency exceeds the capabilities of any real-time (nonsampling) oscilloscope, the sampling scope is the instrument to use.

Storage Oscilloscope

At the opposite end of the spectrum are the signals that occur either too slowly or too infrequently to be recorded by

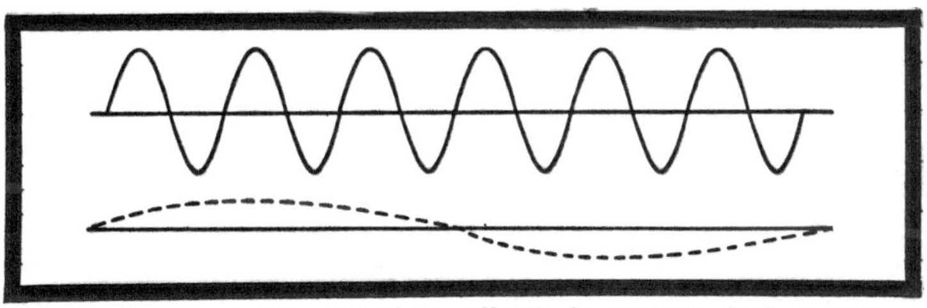

Fig. 4-8. The sampling technique effectively converts high-speed repetitive waveforms into low-speed signals that may be handled by a conventional oscilloscope. In the actual presentation the sampling dots will be much closer together, and the displayed waveform will appear much as shown by the dotted line.

the rapidly fading phospohors of a conventional CRT. The solution to this problem is again a specialized oscilloscope—the storage oscilloscope. The storage oscilloscope is either built around a specialized CRT or digital storage circuitry. In the first case, once the signal has been recorded on the CRT it will remain there permanently until erased. With the digital storage technique the waveform is applied to an A–D converter, and the digitized waveform is stored in memory; the memory is then cycled in step with the time base of a conventional oscilloscope. The output of the memory is applied to a D–A converter, whose output is connected to the vertical-deflection plates of the oscilloscope. Slowly varying signals are speeded up considerably, and signals occuring only once are displayed over and over again. At present, the storage oscilloscope employing the storage CRT has the highest frequency capabilities, which is important in transient (single-occurrence) measurements, while the digital storage oscilloscope has the advantage of indefinite storage and a simple display mechanism. Both are reserved for specialized applications and are expensive.

Television Oscilloscopes

Television signals are complex, high-frequency waveforms. These signals can not be analyzed easily with any other instrument than the oscilloscope. Because of the great need, two specialized forms of oscilloscopes have been developed for the television industry.

The television *waveform monitor*, or *A-scope*, is simply a conventional oscilloscope with limited time bases. Time base periods are limited to those needed to display one or two TV lines and one or two TV fields. Such oscilloscopes are usually mounted in a specific location to monitor the quality of video waveforms from a camera, network, film chain, transmitter, etc.

The *vector monitor*, or *vectorscope*, is an oscilloscope without sweep circuits. Red-chroma-minus-luminance (*R-Y*) signals are connected to the vertical-deflection system, and blue-chroma-minus-luminance (*B-Y*) signals are connected to the horizontal-deflection system. The patterns displayed on the CRT reveal information concerning phase and amplitude characteristics of the chroma signals. The vectorscope is also used by television servicemen.

Developed from the generalized oscilloscopes are many other dedicated (special-purpose) oscilloscopes. Dedicated oscilloscopes have their vertical amplifiers, horizontal amplifiers, and time bases customized to a particular situation. However, one can easily apply the fundamentals of the general-purpose oscilloscope when dealing with one of these specialized instruments. The fundamentals of any particular section of the circuitry are the same.

SPECIFICATIONS

To completely understand the oscilloscope, the exact meaning of the numerous specifications applied to its capabilities must be thoroughly understood. The modern high-performance oscilloscope has many involved and interrelated specifications. It is extremely important when selecting an oscilloscope that the measurement requirements be understood and that all specifications of the instruments under consideration be compared on an equal basis.

VERTICAL SPECIFICATIONS

Vertical bandwidth is one of the most fundamental specifications of an oscilloscope. This specification, more than any other, will determine the suitability of a particular oscilloscope for the job at hand. Oscilloscope manufacturers try to create a vertical amplifier whose frequency response is constant until an upper frequency limit is reached, where a controlled rolloff (decrease in gain) starts. The bandwidth of the oscilloscope is defined as the point at which the displayed vertical signal has been reduced by 3 dB with respect to some low-frequency reference point. As vertical signals continue higher in frequency, the oscilloscope response should continue to roll off at a rate slightly greater than 6 dB per octave. This controlled rolloff is necessary to provide proper vertical-amplifier response to complex signals.

Oscilloscopes having a vertical-frequency response which rolls off at a rate considerably greater than 6 dB per octave will not faithfully reproduce the high-frequency components of complex waveforms. On the other hand, oscilloscopes with insufficient high-frequency attenuation will tend to create *overshoot*, or excessive high-frequency response on pulse signals.

Note that a signal reduced by 3 dB is at its half-power point, not its half-voltage point. At −3 dB the voltage is 70.7% of the reference value. Furthermore, a signal reduced in amplitude by 3 dB because of an increase in frequency has a large phase shift with respect to the reference point, normally in the area of 45°.

Occasionally the vertical bandwidth of an oscilloscope is specified with a deflection limitation. Such a specification might read: 10 MHz at 4 cm deflection, 8 MHz at full deflection. This specification indicates that the oscilloscope may not be used at its full bandwidth if full deflection must be utilized. This specification is popular with some solid-state oscilloscopes, which may have a limited vertical-plate driving capability. Generally speaking, most higher cost modern oscilloscopes are not specified in this manner. However, one should watch out for a deflection limitation when purchasing a unit if this would result in application limitations.

Oscilloscopes come with vertical amplifiers that are only AC coupled and with vertical amplifiers having both AC and DC coupling, usually switch selectable. The AC–DC coupling is the more versatile, but AC-only is generally lower in cost. When an oscilloscope is operated in AC-coupled mode, it will have an upper −3 dB limit set by the vertical-amplifier high-frequency rolloff and a lower −3 dB limit set by the low-frequency reactance of the AC input coupling capacitor. The 3 dB lower frequency limit is usually 2–10 Hz. With AC coupling, the highest potential that may be applied across the input coupling capacitor must be specified. This is usually 400V–600V. This specification is peak AC plus DC, *not* just DC.

The input mode selector may also have a third position in addition to the AC and DC positions described above (Fig. 4-9). This third position, usually described as GROUND, disconnects the input connector from the input amplifier; the input to the vertical amplifier is grounded. This feature is frequently used to note the 0V height of the CRT trace.

Rise Time

Closely related to vertical bandwidth is vertical rise time. Rise time is defined as the time required for the signal to

Fig. 4-9. A closeup photograph of the vertical input of a Heath 10-4510 dual-trace 15 MHz oscilloscope. Notice the 3-position input-coupling switch, with the center decoupling position marked GROUND.

increase in amplitude from 10% of its total value to 90% of its total value (Fig. 4-10). The rise-time of an oscilloscope is important in determining the limits of rise-time measurements that maybe made by the oscilloscope. Rise-time t_r (in microseconds) is directly related to bandwidth f_{MHz} by the formula

$$t_r = 0.35/f_{MHz}$$

An oscilloscope which meets a given rise-time requirement will have proper high-frequency rolloff.

Rise time is especially important if the oscilloscope is to be used for pulse analysis. Then the oscilloscope should ideally have a rise time that is 20% of the value of the pulse to be measured.

Deflection Sensitivity

Vertical-deflection sensitivity ranks equally with vertical bandwidth as an important oscilloscope specification. Both of these limitations can prevent a measurement from being made.

The deflection or input sensitivity specification indicates the smallest voltage that will produce a standard deflection, usually one vertical division on the CRT. Stated in another way, deflection sensitivity indicates the maximum vertical amplification available. As a typical example, an oscilloscope might have a sensitivity of 10 mV. This oscilloscope would display a 10 mV peak-to-peak specification, not RMS. A 10 mV RMS sine wave would cover approximately 2.82 divisions, as a 10 mV RMS sine-wave signal is a 2.82 mV peak-to-peak signal. Sensitivity obviously costs money; therefore, the amount of sensitivity required must be weighed against the cost of the oscilloscope.

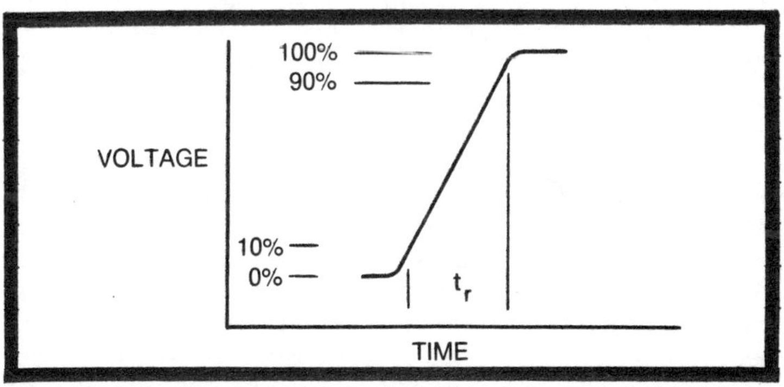

Fig. 4-10. A drawing depicting the definition of rise time. Fall time is defined in a similar manner.

Often the oscilloscope is used with an accessory probe which acts as a voltage divider, reducing the input signal by a factor of 10 or more. If this is the case, one must remember that the osilloscope sensitivity is effectively reduced by the same factor.

To obtain maximum sensitivity, some models offer increased sensitivity at a reduced bandwidth. Oscilloscopes offering this feature typically give an additional gain of 10 with a bandwidth reduced by a factor of 4. For example, a 10 MHz oscilloscope might maintain 10 MHz bandwidth at 10 mV per centimeter, however, the input attenuator may be adjustable to 1 mV per centimeter, but with a bandwidth of only 2.5 MHz. Often this bandwidth is quite different, as high-frequency signals may not be of interest at high-sensitivity levels.

Input Attenuator

The maximum sensitivity of an oscilloscope cannot be utilized on all measurements. For example, in a 10 mV per centimeter oscilloscope with 6 cm of total vertical display, the electron beam will go off the screen for signals in excess of 60 mV. To overcome this problem, an input attenuator is provided. This is usually one of two types. On the simplest oscilloscopes it may be nothing more than a variable control, or at best, a 3-position switch labeled ×1, ×10, ×100. In such oscilloscopes the amount of attenuation and the vertical sensitivity are generally uncalibrated. The more sophisticated oscilloscopes have an attenuator with steps calibrated in the resultant vertical deflection sensitivity. This is usually in a 1—2—5 sequence, although occasionally a 1—3—10 sequence is used.

For example, the oscilloscope shown in Fig. 4-9 has a basic deflection sensitivity of 1 mV/cm. Because of high impedances and stray capacitances, a simple resistance divider will not maintain the same attenuation at high frequencies as at DC. To correct this, the input attenuator must be capacitively compensated. With compensated attenuators the attenuation is constant at all frequencies.

The step attenuator has a disadvantage in that it will not allow signals of any arbitrary amplitude to be made exactly full scale or some other desired size. To permit such operation, most oscilloscopes include a variable control that adjusts the effective attenuation between the indicated value and its next highest position. For example, an oscilloscope used at 500 mV per divison can be adjusted continuously between 500 mV per division and 1V (1000 mV) per division by this control. This variable control will have a CALIBRATED position (normally extreme clockwise). In the CALIBRATED position, the deflection factors indicated on the attenuator positions fall within the accuracy limits set for the oscilloscope. Accuracy of attenuation is normally ±3% to ±5%. Vertical-accuracy specifications also include any inaccuracies found in the vertical amplifier. Accuracies are frequently not specified at the high-frequency limits, at temperature extremes, nor on extremely low cost instruments.

Input Impedance

For most servicing and development work, a high input impedance is desirable. One megohm has been chosen as a standard. As was noted, there is compensating capacitance involved with attenuators. Therefore, the input impedance of an oscilloscope includes the value of capacitance found in parallel with the 1M resistance. This capacitance usually lies in the area of 20–40 pF if the oscilloscope is designed to be used with a divider probe. Obviously, the lower the capacitance the better. Other impedances have been used. Some of the older, very low cost oscilloscopes have inputs ranging from 100K to 10M. Some of the very sophisticated high-frequency oscilloscopes built today have a nonreactive input.

Input Connectors

Most oscilloscopes are provided with a BNC input connector (see Fig. 4-11). This is the most desirable, considering the availability of cables and probes with mating BNC connectors. Other input connectors used are the 3- or 5-way binding posts, generally used on very low cost oscilloscopes and the UHF connector (SO-239), found on some older models.

Positioning Range

Vertical amplifiers are provided with a continuously variable control permitting the operator to adjust the vertical position of the trace. The range of the POSITION control and the effect of extreme positioning will be different for different oscilloscopes. The range of the POSITION control is measured in *windows*. A window is the full distance across the CRT in the direction in question, in this case, the usable vertical dimension of the CRT. An oscilloscope specified as having two vertical windows is capable of deflecting a waveform occupying the full vertical display area upward to the extent that the bottom of the waveform is at or above the display centerline. The control should also be able to deflect the trace

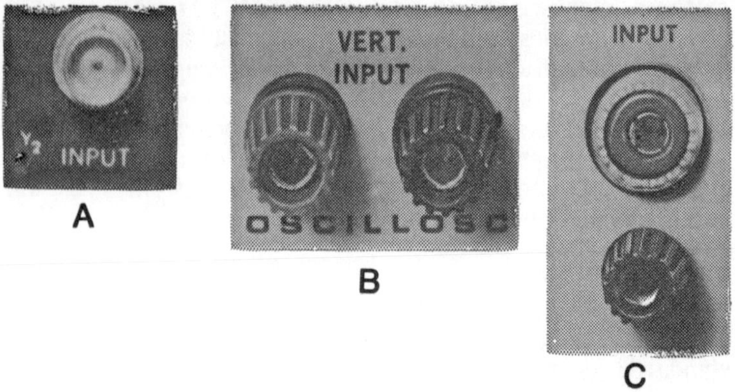

Fig. 4-11. Common input connectors used for oscilloscopes: (A) BNC, (B) 5-way, and (C) UHF. Most frequently the vertical input will be one of the coaxial types; however, the horizontal and trigger inputs may be 5-way connectors.

downward until the top of the waveform will be at or below the centerline. When the POSITION control is adjusted to either of these extremes, there should be no on-screen distortion of the waveform. The POSITION control may have more range, but beyond the two windows, there may be distortion of the trace. Vacuum-tube oscilloscopes tend to have a larger number of vertical windows. The more modern solid-state oscilloscopes are frequently limited to two windows, and many have but 1½ windows.

Vertical-Delay Lines

To allow for trigger and sweep startup, a delay line is inserted in the vertical-amplifier circuit. The object of the delay line is to uniformly delay signals of all frequencies by an amount slightly greater than the time required to permit triggering, startup of the sweep circuits, and an unblanking of the tube before the triggering signal is presented to the CRT. Specifications will indicate the amount (number of nanoseconds) of pretriggered waveform which will be displayed.

Delay lines are essential for good pulse analysis. They are especially important with digital circuitry, in which the measurement of pulse rise times may be critical to proper operation.

TIME BASE

As noted earlier, recurrent sweep is the simplest form of time base available. The recurrent sweep time base offers no way to make calibrated time measurements except by comparison. The recurrent-sweep-generator specifications indicate the upper and lower frequencies of the sweep oscillator. The frequency can be changed with a variable control within a decade range, and over multiple decades in switched steps. Sweep oscillator frequency range from 5 Hz to 500 kHz is typical. Converted to time per division, assuming 10 horizontal divisions, this gives an equivalent range of 20 msec per division to 200 nsec per division. A few oscilloscopes have provisions to lower the sweep oscillator frequency by use of an external capacitor.

The recurrent sweep time base may also have a control to adjust the amplitude of the synchronizing signal injected into the sweep oscillator from the vertical amplifier. Switch selection of a positive- or negative-going synchronizing signal is frequently made available. Often this same switch will

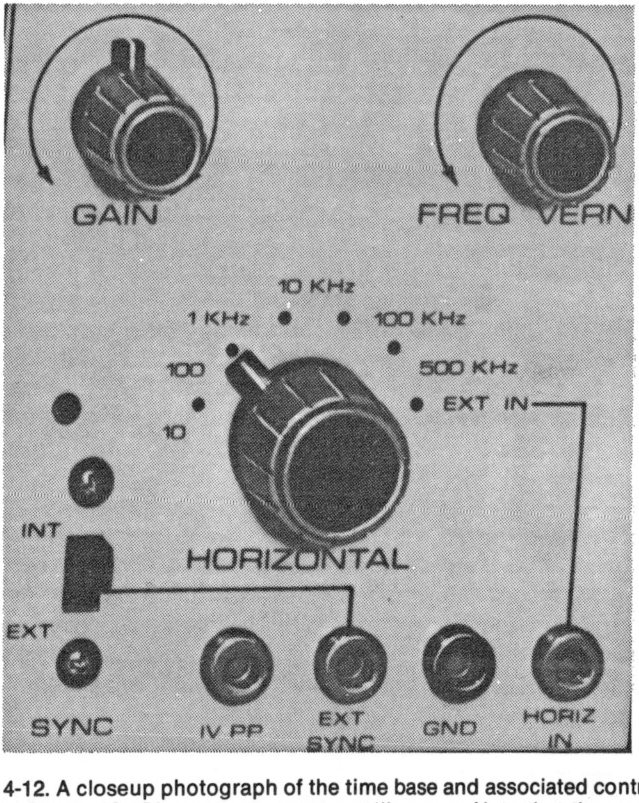

Fig. 4-12. A closeup photograph of the time base and associated controls on the Heath 10-102 recurrent sweep oscilloscope. Note that the coarse frequency control is also used to select external horizontal inputs.

permit positive or negative synchronizaton by an external signal or a sample of the power-line frequency. (See Fig. 4-12).

Calibrated Sweep

Calibrated sweep yields a time measurement device. The period of the time base is selected as a time per division.[1] The switching is either in decades (on lower cost oscilloscopes) or a 1–2–5 sequence.

Slowest sweep speeds vary with the manufacturer and the price of the oscilloscope, but usually they are in the vicinity of 200 msec per division to 2 sec per division.

The fastest sweep speed is dependent upon the bandwidth limit of the oscilloscope. A rule of thumb is that the fastest sweep speed should present no less than three complete cycles of a waveform whose frequency is identical to the vertical bandwidth of the oscilloscope. For example, a 10 MHz oscilloscope would require an upper sweep speed $3 \times 1/(10 \times 10^6)$ sec = 300 nsec for the full horizontal span, or 30 nsec per division. This requirement would be met by an oscilloscope time base having a maximum speed of 200 nsec per division and a ×10 magnifier yielding a 20 nsec per division display.

For low-bandwidth oscilloscopes (3–5 MHz), fastest sweep speeds in the area of 0.5–1 μsec are typical. Sweep speeds of 0.05–0.2 μsec per division are common on 50 MHz oscilloscopes.

Although the time per division may be stepped in either a 1–2–5 or 1–10–100 sequence by the time base switch, there is usually provision to vary the time per division continuously between steps with an uncalibrated control. Accuracy of the time base is usually ±5% to ±3%. Time base speeds are often affected by temperature, line voltage variations, and age, so they should not be used as ultimate standards of time comparison.

Triggering Controls

The triggered oscilloscope gains much of its flexibility from the various modes of operation which may be selected for the trigger circuits (Fig. 4-13). The trigger signal, taken from the vertical amplifier, is used to start the sweep generator. Variations of triggering include selection of positive or negative triggering, level of triggering, AC or DC coupling, high- or low-frequency filtering, and selection of trigger source

---

[1] The majority of oscilloscopes use the centimeter as the basic horizontal and vertical division. There are some oscilloscopes with vertical divisions slightly longer or shorter than a centimeter. For this reason the general term *division* will be used.

(including an external source, the power line, and the vertical-amplifier channels). Each of these features adds to the ability to observe complex waveforms.

Fig. 4-13. A closeup photograph of the triggering and sweep controls of the Heath 10-4510 dual-trace, 15 MHz oscilloscope. Note that in addition to source and slope selection, the operator may select DC, AC, or ACF (AC coupled through a 15 kHz high-pass filter) coupling. The auto mode is selected by full counterclockwise operation of the TRIGGER LEVEL control.

Older oscilloscope designs also incorporate a STABILITY control which assists in proper operation of the trigger circuits. The STABILITY control is adjusted prior to using the TRIGGER LEVEL control. Generally speaking, STABILITY controls are not found on modern oscilloscopes.

The methods of defining trigger sensitivity and triggering bandwidth are not consistent. The following are Heath Company standards for such measurements. *Trigger sensitivity* indicates the smallest deflection (or external input level) that will permit a stable trace on the face of the CRT. Sensitivity of one division or less is desirable. An oscilloscope requiring more than one division of vertical deflection to maintain a stable display does not have sufficient trigger sensitivity for many applications.

Trigger bandwidth can be defined as the highest frequency at which a stable trace can be maintained with some nominal deflection (often one division). Trigger bandwidth determines the ease with which the oscilloscope will trigger on complex waveforms and what the stability of high-frequency signals will be. An oscilloscope with a trigger bandwidth twice the vertical bandwidth provides exceptional triggering performance, while one with a triggering bandwidth of less than its own vertical bandwidth creates difficulties when high-frequency or complex waveforms are being observed.

Time Base Modes

The time base generator usually has two modes of operation. In the *normal* mode the sweep generator is cycled by a trigger pulse, which follows the sweep and holdoff periods. In the *automatic* (*auto*) mode the oscilloscope automatically generates trigger pulses in the absence of a signal in the vertical amplifier. This provides baseline (trace) in the absence of a vertical signal rather than the blank CRT seen in the *normal* mode.

Some oscilloscopes provide a time base mode called *single sweep*. This permits the operator to select a set of conditions that will trigger the sweep and then to "arm" the sweep. When the set of conditions occurs, the time base will be activated for one sweep and then will remain locked out until rearmed. This mode is especially useful when attempting to observe fast events occurring randomly and at widespread intervals. Frequently such events will be detected by an oscilloscope camera.

HORIZONTAL SPECIFICATIONS

Although the primary source of horizontal signals is the sweep circuit of the oscilloscope, there are occasions requiring the horizontal deflection signal to be derived from an external source.

Horizontal Bandwidth

As the only requirement of the horizontal amplifier is to pass the sweep signals with reasonable fidelity, the bandwidth of this amplifier is generally not great. Bandwidths of the order of 1–3 MHz are quite common. Occasionally a high-frequency signal must be handled, and at such times the horizontal bandwidth of the oscilloscope must be taken into consideration.

One should also remember that when specifying a limit of horizontal bandwidth, such as −3 dB at 3 MHz, the manufacturer is also specifying a phase shift. In certain measurements (especially phase measurement), phase shift in the horizontal amplifier can cause measurement errors.

External Horizontal Input

The input impedance of a horizontal amplifier varies from oscilloscope to oscilloscope. However, for most oscilloscopes it is specified as 100K or 1M with some shunt capacitance. On more elaborate oscilloscopes, the horizontal-sensitivity specification may also include specifications for a horizontal attenuator and a variable gain control. The most limited oscilloscopes have only a fixed amplitude specified for horizontal sensitivity. External horizontal-input connectors will normally be the same as those for the vertical input. However, the 5-way binding post is used occasionally when the vertical-input connector is a BNC type.

OPERATIONAL FEATURES

Previous discussions have placed specific numbers on operational characteristics of oscilloscopes. A group of oscilloscope characteristics equally important but without numerical specifications are the oscilloscope features. The oscilloscope is an extremely complex instrument and a highly versatile tool. It is the operational features of the oscilloscope that often give it the greatest possible usefulness to the engineer or technician. A thorough knowledge of the available features and how they assist in certain measurements is valuable for selecting and using an oscilloscope.

Vertical Modes

The vertical modes of an oscilloscope can vary from the simple situation of one mode, the single input on a single-trace oscilloscope, to the many modes available on some models of dual-trace oscilloscopes. The simplest dual-trace oscilloscope will have at least four modes of operation: channel 1, channel 2, channel 1 chopped with channel 2, and channel 1 alternated with channel 2. Vertical modes may be selected by a VERTICAL MODE switch, or the modes may be selected by a combination of the TIME BASE switch and the POSITION controls. In the latter case, each POSITION control has a switch position at one extreme that will turn off the trace. If both traces are turned on, then chopped or alternate mode is selected by the TIME BASE switch. This method of vertical-mode selection does not allow the user to select when he will change from chopped to alternate mode, but it does eliminate operator errors caused by failure to remember to switch from one mode to another.

On the more sophisticated oscilloscopes an additional mode is added, *algebraic add*. When the VERTICAL MODE switch is at ALGEBRAIC ADD, signals from channel 1 and channel 2 are added together and displayed. This mode lets the operator compare two signals. Often the ALGEBRAIC ADD switch is accompanied by a switch to invert the polarity of one or more of the vertical channels. When operated in this mode, with one channel inverted, signals common to both inputs will not be displayed, as they subtract. However, signals not common to both of the inputs will be displayed. The ability of most oscilloscopes to operate in this mode is confined to relatively low frequency signals (a few hundred kilohertz and below) and to signal amplitudes within a few vertical windows.

Calibrators

Voltage calibrators are generally found on all high-priced oscilloscopes, and to varying degrees and with varying quality, they are found on low-priced oscilloscopes.

A voltage calibrator which provides a square wave of reasonable rise time (at least a microsecond or two) and a calibrated voltage waveform serves a number of purposes. It permits the user to check the amplitude accuracy of the vertical amplifier when he desires. It permits him to establish a gain level using the variable vertical-gain control. And it permits compensation of oscilloscope probes to insure the frequency response of the probe is matched to the particular oscilloscope input.

On the simpler oscilloscopes, the only waveform made available may be a sinusodial waveform with known vertical amplitude. In extremely simple oscilloscopes that only have a simple vertical attenuator and an uncalibrated amplifier, this may be used to establish the amplitude of the measured waveform by relative measurements. Outside voltage sources may be used to calibrate the oscilloscope; however, the vertical calibrator signal is always a desirable feature when it can be obtained.

CRT

A limitation on the high-frequency capability of the oscilloscope, above that of the bandwidth of the vertical amplifier, is the ability of the CRT to write (produce traces) at high speeds. That is, the phosphor must be sufficiently sensitive and the electron beam sufficiently intense to cause the phosphors to emit light in the time the beam strikes them. At higher and higher rates of trace deflection, this becomes more and more difficult, until a point is reached where the CRT is unable to display the signals in the vertical amplifier.

Another situation causing this same condition is a moderate to fast sweep speed, with infrequent sweep triggering caused by the low rate of the desired waveform. An example of this is a 60 Hz waveform being observed at a 10 μsec/cm setting. In such a case the phosphors are being refreshed only infrequently (once every 16 msec), and unless the oscilloscope CRT has good writing speed, the trace will be very faint. Unfortunately, the writing speed of an oscilloscope is not easily specified. Generally speaking, the higher the accelerating potential, the better the writing speed.

Acceleration

Two different types of CRTs are available, and these should be watched closely. The simplest is what is known as a *monoaccelerated* tube, in which all the electron beam acceleration takes place within the gun assembly before the deflection plates, usually with voltages of 2 kV or less. The most common circuits use 1.4 kV acceleration.

Postdeflection Accelerator

In more sophisticated oscilloscopes, a more expensive tube with *postdeflection acceleration* (PDA) is used. Tubes with PDA will generally exhibit much better writing characteristics as much higher voltages are used, generally between 4 and 20 kV. Generally speaking, oscilloscopes with a vertical bandwidth exceeding 10 MHz should use a PDA tube for optimum results.

Beam Controls

The electron beam in the CRT is focused, shaped, and accelerated by various elements within the CRT. A minimum of two of these elements have controls on the front panel of the oscilloscope. These controls permit operator adjustment of the intensity and focus of the beam. An ASTIGMATISM control is often made available to shape the spot. The ASTIGMATISM control may be located on the front panel, the rear panel, or sometimes inside the case. The FOCUS, INTENSITY and ASTIGMATISM controls interact to a considerable degree.

Rectangular and Flat Tubes

The original oscilloscope tube looks somewhat like a wine bottle. This tube has served for many years, despite the generally preferred viewing area being a rectangle. With the round-face tube, a portion of the corner display area is lost (see Fig. 4-14), and there is a large unused portion of the screen. While many oscilloscopes today still have round tubes, they are frequently masked off to show a rectangular viewing area. Generally speaking, this is of no detriment to the oscilloscope user. The rectangular tube also makes more efficient use of the front-panel space.

A rectangular tube requires a control for trace rotation. This control—which is not needed on the round tube, as it may

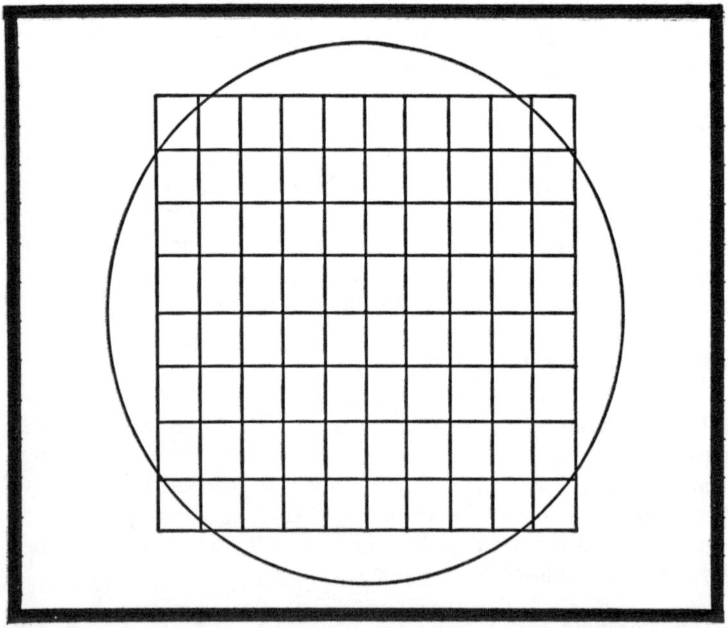

Fig. 4-14. A maximum-sized rectangular graticule placed on a round-faced CRT. Note that the outermost squares at each corner are lost due to the round edge of the tube. A 5 in. round tube will give an 8×10 cm graticule. Although some squares are lost, these are rarely needed to make a usable display.

be simply rotated to produce a horizontal trace—rotates the electron beam with a magnetic field produced by a DC current in a coil wrapped around the tube. True rectangular tubes are generally of the PDA type and represent a significant increase in construction cost.

The original oscilloscope tube had a slightly convex display area. This was necessary so the beam would remain focused on the face of the CRT at any position. A convex tube face leads to errors due to parallax. So this CRT has been replaced by the flat-face tube. The flat-face tube has its focusing problems corrected by the internal mechanics of the CRT and eliminates the parallax error.

Viewing Area

The viewing area of the CRT is generally divided into squares, with the most common being 8 × 10 or 6 × 10 cm. However, some oscilloscopes use smaller units of measurements, sometimes as small as 0.6 or 0.8 cm per division; and some oscilloscopes use larger divisions, such as 1.2 cm per division. Generally speaking, the larger the screen size, the more easily the oscilloscope is read.

The extremely small tubes, those using 0.6 and 0.8 cm per division, were popular with some of the early portable solid-state oscilloscopes, in which it was difficult to obtain large deflection voltages. The 8 × 10 cm screen is by far the most popular today, although the 6 × 10 cm screen is also common. Generally speaking, the 8 × 10 screen with 1.2 cm per division is only available on high-priced oscilloscopes. On some low-priced oscilloscopes the diameter of the tube is given. This is usually 5 in. or 3 in. A 3 in. tube will give a small display and should be considered only when small size or portability is a necessity. A 5 in. diameter tube will give an 8 × 10 cm rectangular display (see Fig. 4-14).

Graticules

The graticule is a thin window placed over the face of the CRT. It bears the lines defining the rectangular divisions. The oscilloscope manufacturer constructs the oscilloscope so the graticule lies as close as possible to the face of the CRT. This is required to minimize the parallax error between the trace and graticule. This problem can be absolutely minimized by etching the lines on the inside of the faceplate of the CRT. Such internal graticule tubes are expensive, and thus this feature is found only on higher priced oscilloscopes.

To make the lines of the graticule stand out, the more expensive oscilloscopes have an illuminated graticule. This can be achieved by simply flooding the graticule with the light generated by a lamp similar to a pilot lamp, or it may be accomplished by grinding the edge of the graticule (or the faceplate of the CRT on an internal-graticule tube) and illuminating this ground edge. If the lines have been scribed in the graticule rather than painted on, extremely uniform illumination of each of the marks and lines will result. If oscilloscope photography is contemplated, a scribed graticule with probe illumination is mandatory. If scale illumination is available, a control is normally provided to vary the intensity.

Phosphors

The phosphor on the inside of the CRT face turns the electron beam into a visible trace. There are three different types of phosphors common to oscilloscopes. Type P1 is the most common phosphor on the low-cost, round CRTs. The P1 phosphor emits a yellow-green light, has a moderate resistance to burns, and has a decay to 0.1% of maximum light output in 95 msec. Wherever possible, the P1 phosphor is being replaced with a newer phosphor, the P31. The P31 phosphor emits a yellow-green light similar to that of the P1 but at nearly twice the intensity. The burn resistance of P31 is considerably better than that of P1, an important consideration as this is a frequent cause of damage. The P31 phosphor will appear to flicker at a much faster sweep repetition rate than will the P1, though, as it has only a 32 msec life of 0.1% output.

The P7 phosphor appears in a few special-purpose oscilloscopes where long persistence is required. Decay to 0.1% output is nearly 1.5 sec, and under low-level lighting conditions, a trace can be seen over a full minute. The P7 phosphor fluoresces with a blue light when excited and then emits a yellow phosphorescent glow. Light output of the P7 is the lowest of all the phosphors, and it can be easily burned.

All phosphors can be damaged by permitting a high-intensity trace to remain in one spot too long. Make a habit of turning down the intensity or of defocusing an oscilloscope when it's left on for long periods of time. A single small spot left on the face of the CRT will damage the phosphor very rapidly, especially if the P7 or P1 type is used.

Regulated Supplies

The accuracy and stability of the oscilloscope are directly related to the stability of its power supplies. The deflection sensitivity of the CRT is directly dependent upon the accelerating potentials supplied to it, and the accuracy of the vertical deflection is directly proportional to the stability of the deflection sensitivity. The accuracy of the oscilloscope thus becomes dependent upon the stability of the high-voltage power supplies. Regulated high-voltage power supplies tend to be expensive, and some manufacturers forgo regulation to produce lower cost oscilloscopes. As a result, traces bounce or jump and reductions in horizontal and vertical accuracy occur with power line variations. A good specification to look for is one indicating the power line voltages over which the oscilloscope may be operated.

XY Mode

An XY mode is a feature occasionally found on single-trace scopes and more frequently found on dual-trace models. Accompanying the XY feature is a specification indicating the degree of differential phase shift between the X and Y channels. This may be something like 3° at 100 kHz. The XY mode is frequently used for making phase measurements. Therefore, with an oscilloscope that has 3° differential phase shift at 100 kHz, a measurement indicating a 45° phase shift at 100 kHz between the signal supplied to the X and Y inputs could be in error by 3°, or 7%. Oscilloscopes with vertical-delay lines present an additional problem, as the delay line itself creates an effective added phase shift and must be compensated for in the horizontal amplifier, or the user must tolerate the added phase shift contributed by the delay line.

Magnifiers

Often it is desirable to observe a signal in more detail. With a triggered oscilloscope this generally is accomplished by increasing the sweep speed. However, there are occasions when a magnifier is used instead. Most magnifiers have magnification factors of 5 or 10, although sometimes a factor of 2 is used. There are two forms of magnifiers, horizontal and sweep.

A horizontal magnifier, by far the more common, extends the length of the sweep by the magnification factor. Thus, a ×10 magnifier effectively makes a 10-division trace 100 divisions long. Using the HORIZONTAL POSITION control, different 10-division sections of the 100-division waveform can be observed. Effectively, the time per division is shortened by a factor of 10, so a 10 μsec per division sweep speed is effectively made 1 μsec per division with a ×10 horizontal magnifier. The trace will tend to dim when a horizontal magnifier is in use, as the beam is on the screen on the CRT only one-tenth as often as it would have been without the use of the magnifier. A sweep magnifier multiplies the repetition rate of the sweep circuit by the magnification factor, and the trace does not dim. The effect, in terms of sweep speed per division, is still the same. As noted before, sweep magnifiers are rather uncommon. Frequently magnifiers contribute additional error, often as much as ±2–3%.

Packaging

Oscilloscopes come packaged in one of three basic styles: portable, bench, plug-in or nonplug-in. Depending on the service required of an oscilloscope, each of these various configurations has advantages. There is no perfect package, and tradeoffs must be made.

The vast majority of oscilloscopes intended for field service fall in the portable classification. Portable implies small size and weight, low profile, and an instrument which is easily carried. More and more oscilloscopes classified as portable have a battery option. At present this is not an absolute requirement for a portable classification. Most oscilloscopes are in the 20W–60W category and require a fairly large and expensive battery pack. Oscilloscopes intended for extensive field use should be fairly rugged and have a wide range of line voltage capability. Line voltage variations from 105V to 130V are not at all uncommon, and field service may well place the oscilloscope at remote points where the line drop is great and regulation is poor. A portable oscilloscope trades compactness for a slightly crowded front panel, a smaller CRT, and an increase in price over a larger, nonplug-in bench equivalent. Portable oscilloscopes are less likely to have all the extras found on a bench instrument.

Most of the older, less expensive instruments fall in the nonplug-in, bench category. These units tend to be taller than they are wide, but they are normally of the same depth as the portable units, depth usually being a function of CRT length. Power requirements may be high on these instruments, and their portabiliy is limited to moving the unit from bench to bench.

The plug-in feature rarely finds its way into the type of oscilloscopes being covered in this book. Plug-in units are designed with interchangeable sections for the vertical preamplifier and sometimes the time base. Plug-in oscilloscopes permit one mainframe to be used with a wide range of inputs and time bases. Their main advantage is in the laboratory, where the mainframe is very expensive because of high-performance requirements, and the extra cost of plug-in capability does not add measurably to the cost of the product.

Most plug-in oscilloscopes are anything but portable—frequently carts are used to move them about.

The features listed previously are not all that are available today. In the more sophisticated oscilloscopes, one of the most common optional feature is a second time base. A second time base permits such operations as *delayed sweep*, wherein the sweep of the CRT is started after a specific period of time, determined by the first time base, has elapsed. *Mixed sweep* permits a portion of the displayed trace—say, the first three divisions—to be made at one sweep speed, and the next three divisions to be made at second sweep speed. Both delayed and mixed sweeps have the effect of acting as very sophisticated magnifiers. They are especially useful when it is not possible to trigger on that portion of the waveform that must be observed.

Other features available in higher priced oscilloscopes include chopped blanking, outputs from many of the sweep and gating circuits, blanking output and input trace intensification inputs, camera mounts permitting photographic work, and optional phosphors.

BASIC OPERATING TECHNIQUES

Basic operaton of most instruments is straightforward, and little knowledge is required of the user when moving from a scope of one manufacturer to that of another or when performing simple tasks. The oscilloscope does not follow this pattern. The many types of oscilloscopes available and the relative complexity of all oscilloscopes in comparison to other instruments make it necessary for the user—especially the casual or infrequent one—to follow some simple rules of initial operation. If these are not followed, considerable time and frustration results from attempting the apparently simplest of procedures. Therefore, the object of this section is to establish initial operating rules for a basic oscilloscope. Once the rules are mastered, the user has a basis for operating a more complicated and sophisticated instrument, such as a triggered oscilloscope.

Obtaining a Trace

When a new or unfamiliar oscilloscope is first operated, the first procedure is to obtain a trace. The oscilloscope should be plugged in, turned on, and permitted to warm up. Some oscilloscopes, especially those using vacuum tubes, require 1 or 2 minutes before any trace can be obtained. All oscilloscopes should be given additional warmup time before use to eliminate trace drift.

After warmup, set the time base (VOLTS—DIV) switch to a medium sweep speed and place the AUTO—NORMAL switch (triggered oscilloscopes) in the AUTO position. Increase the INTENSITY control, with the VERTICAL and HORIZONTAL POSITION controls set at the center of rotation, and a trace (horizontal line) will appear. If at full intensity there is no trace, look for a general lighting at one of the edges. This gives an indication of the off-screen direction of the trace. The VERTICAL and HORIZONTAL POSITION controls can be used to return the trace to the centerscreen. If this technique is not successful, a more systematic approach may be used. Return the VERTICAL and HORIZONTAL POSITION controls to their midway points. Now systematically vary the HORIZONTAL POSITION control in small amounts. After each correction of this control, operate the VERTICAL POSITION control through its entire range to locate the trace. Do not vary the VERTICAL POSITION control quickly, or the beam will cross the CRT face too rapidly to be seen.

Beam Adjstments

Once a trace has been found, change the oscilloscope time base to an EXTERNAL horizontal position. A dot rather than a line will now appear on the face of the tube. Now adjust the FOCUS, INTENSITY, and ASTIGMATISM controls to obtain the smallest, roundest dot possible. This procedure insures the sharpest trace. The FOCUS, INTENSITY, and ASTIGMATISM controls all interact. Trace brightness variations may require refocusing.

Displaying A Waveform

The VERTICAL and HORIZONTAL POSITION controls are adjusted to center the trace vertically and to start the trace at the left side of the display. The time base should be at midrange speed, set for internal trigger or synchronization (either + or −), and a triggered oscilloscope should be in the *automatic* mode.

Once a trace is obtained, a signal must be connected to the oscilloscope to observe a waveform. Obviously some form of cable or probe must be used to connect the signal to the VERTICAL INPUT jack. The VERTICAL INPUT switch is then used to select the AC or DC position, depending upon the information desired from the trace. A common error is to leave the VERTICAL INPUT selector AC—DC—GROUND switch in the GROUND position. If uncertain about the relative amplitudes of AC and DC components of the input waveform, the AC position should be used first. An attenuator setting commensurate with the input signal amplitude should be selected. If the amplitude is unknown, the *highest* attenuation level (lowest sensitivity) possible should be selected.

The attenuator is adjused to produce a trace from one-half to three-fourths of screen height. When performing initial adjustments, both the vertical attenuator and the horizontal variable controls should be in their calibrated positions. Leaving these controls in their uncalibrated positions is a frequent source of error.

Once a trace has been obtained, the VOLTS—DIV switch should be adjusted to display a few cycles of the input waveform. The trace may not be stable at this time; however, usually a sufficient amount of the unstabilized trace may be observed to yield an approximate time base setting. With this setting, a triggered oscilloscope in the AUTO position may require adjustment of the LEVEL control to produce a stable pattern. Some oscilloscopes utilize the LEVEL control in the AUTO position, and some do not. With the recurrent sweep oscilloscope, the COARSE and FINE time base controls are adjusted to produce the desired number of cycles of display; if a SYNCHRONIZATION LEVEL control exists, it may need adjustment to stabilize the trace.

Triggering in the Automatic Mode

The *automatic* (*auto*) mode of triggering, *autotrigger*, is usually the most frequently used. Autotriggering has the advantage of maintaining a trace (baseline) when triggering is lost due to insufficient signal.

Autotrigger is designed to trigger relatively simple waveforms above 100 Hz to an upper frequency limit near that of the vertical bandwidth of the oscilloscope. If the frequency of the waveform is out of this range or the wave is complex, *normal* mode is preferred.

In the ac-coupled *auto* mode, a trigger signal is generated when the displayed waveform passes through the 0V point of the AC portion of the waveform. AC coupling should normally be used, as the *auto* mode is not designed for low-frequency work. Below the frequency which is the autotrigger rate of the baseline generator, the automatic baseline generator takes over between cycles of the displayed waveform. When this occurs, *normal* mode should be used.

When DC-coupled automatic triggering is used (not available on all oscilloscopes), a triggering signal is generated as the displayed waveform passes through an imaginary line representing 0V. The user must be sure that the signal does pass through this 0V line to achieve triggering. To determine where this imaginary line is, change the VERTICAL INPUT selector switch to the GROUND position. This operates the *auto* baseline and establishes a trace whose position is the 0V imaginary line. Varying the VERTICAL POSITION control varies this line, which represents the trigger point. DC coupling of the *auto* mode may be necessitated by signals of low duty cycle or those with a DC offset greater than their amplitude. Figure 4-15 shows the effect of varying the VERTICAL POSITION control with AC and DC trigger coupling in the *auto* mode.

If AC–DC trigger coupling is not available on a particular oscilloscope and DC coupling is necessary to properly trigger the waveform, a change to *normal* triggering is indicated. Such oscilloscopes treat *auto* triggering as AC coupled and *normal* triggering as DC coupled.

Slope Control

The SLOPE control permits the user to select triggering on either the positive- or negative-going slope of the waveform. Figure 4-16 shows the effects of slope adjustment. One of the most important uses of slope selection is in the examination of pulses in which the desired point for observation is on one edge or another of the wave. Slope selection, followed by trace expansion with the VOLTS–DIV control, will permit detailed examination as shown in Fig. 4-16. Slope selection is independent of trigger mode or coupling.

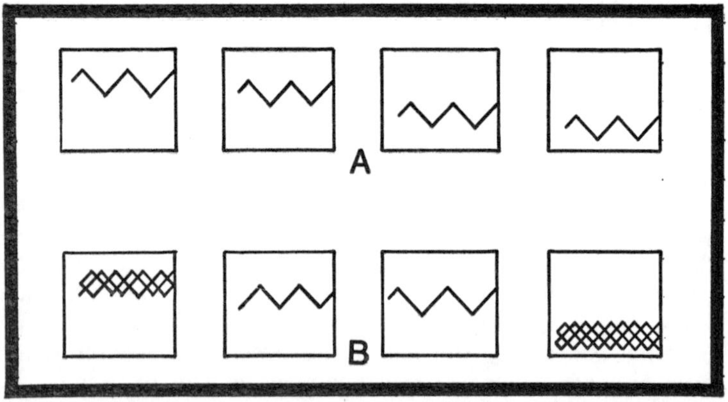

Fig. 4-15. The effect of DC coupling on a trigger signal. The waveforms of section A have AC-coupled trigger signals, and the triggering takes place as the waveform itself passes through its own zero point. The waveforms of section B have DC triggering. Triggering only takes place when the signal passes through the center of the screen (previously established at 0V). Note that when no signal passes through the zero line, a complete loss of triggering occurs.

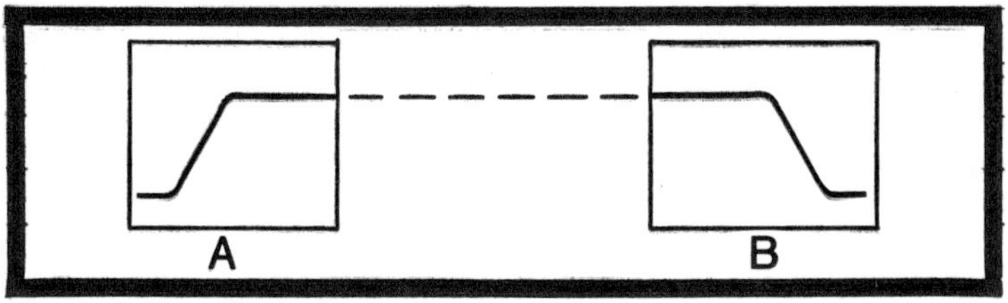

Fig. 4-16. The effect of actuating the SLOPE switch when observing the leading and trailing edges of a pulse whose duration is greater than the time represented by the VOLTS/DIV and the available number of divisions on the oscilloscope screen. (A) Triggering is selected on the leading or + edge. Note some of the pretriggered waveform is visible, indicating the presence of vertical-delay lines. (B) Triggering is selected on the trailing or − edge of the pulse.

Triggering in the Normal Mode

Normal-mode triggering is useful when triggering at a specific point on the displayed waveform is desired. When this mode is employed, the LEVEL control is used, as well as the COUPLING and SLOPE controls. Figure 4-17 shows the effect on a triangle wave of varying the LEVEL CONTROL. The *normal* mode is needed only when the signal cannot be properly triggered in the *auto* mode. An oscilloscope being used at the limits of its frequency capability requires the *normal* mode to secure good triggering.

At the opposite end of the frequency spectrum, this mode is also required. As noted above, the automatic baseline generator may interfere with low-frequency triggering in *auto* mode, and on some oscilloscopes there is no capability to truly DC-couple the trigger signal in the *auto* mode. In this case, signals that are extremely low in frequency or those having DC offset greater than their AC amplitude must be triggered in the *normal* mode. This is especially true of older oscilloscopes.

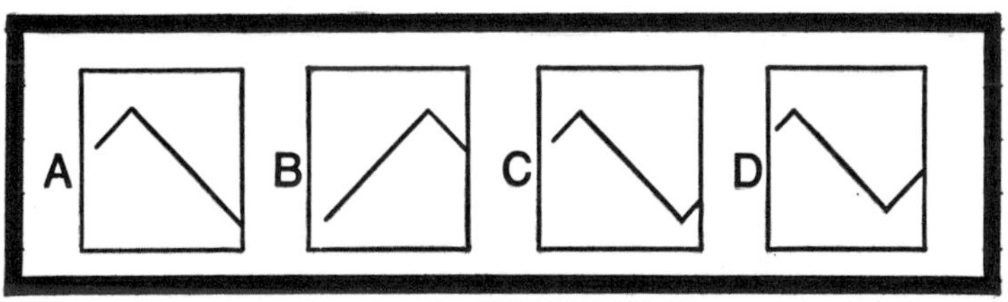

Fig. 4-17. The effect of varying the TRIGGER LEVEL control while observing a triangle wave triggered on the leading (+) edge. (A) The LEVEL control is in the center of its rotation. Triggering occurs near the triangle wave's zero-crossing point. (B) The LEVEL control is close to the CCW point, where triggering will cease. Triggering at the positive or negative turnaround point on a waveform is quite difficult. (C, D) Triggering at greater and greater CW positions of the LEVEL control from the position displayed at A.

In some of the most modern oscilloscopes, the difference between the modes is simply one of an automatic baseline generator. In these instruments, the reason for use of *normal* mode may be no more than keeping a trace off the CRT when no signal is applied or operating at very high frequencies.

Vertical Measurements

The simplest of all vertical measurements is that of peak-to-peak amplitude. Figure 4-18 shows a waveform with a 3 cm vertical amplitude. The vertical attenuator is set at 0.2V per cm to produce this display; therefore, the waveform amplitude is 0.6V (3 cm × 0.2V/cm). It should be remembered that the measurements made are peak-to-peak voltage measurements. An oscilloscope cannot directly make RMS measurements. The RMS value must be calculated by dividing the peak-to-peak amplitude by 2.828. For the square wave, the peak-to-peak value *is* its RMS value. For complex waveforms, conversions involve extensive calculations.

If the oscilloscope has DC coupling, measurements of DC voltage can be made. A measurement of a DC plus AC can be made. Figure 4-19A shows AC vertical coupling with the waveform centered on the screen. The oscilloscope is set to 10 mV per division. A sine wave is measured at 80 mV peak-to-peak. In Fig. 4-19B the input attenuator reduces the vertical sensitivity of the oscilloscope to 1V per division. With DC coupling, the AC waveform is seen to be riding on a DC signal of 3V. Using the GROUND position of the vertical-input coupling switch is a common way of establishing a 0V reference line.

Horizontal Measurements

While vertical measurements yield signal amplitude, horizontal measurements give a measurement of time. Most triggered oscilloscopes with calibrated time bases specify the time per division. As most measurements involve more than

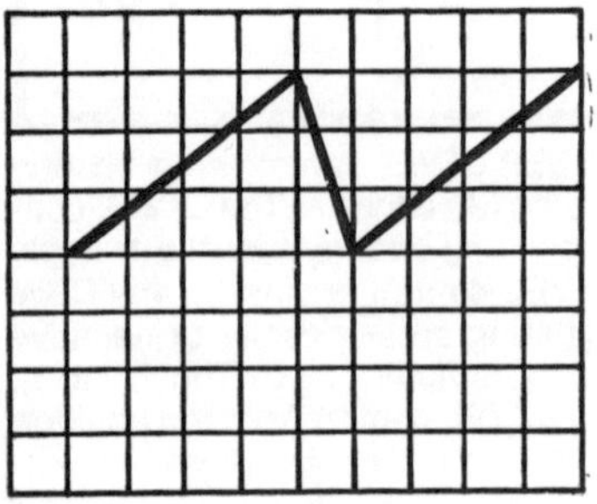

Fig. 4-18. A ramp signal displayed on an oscilloscope with a calibrated-vertical-attenuator setting of 0.2V per division. The peak ramp voltage is 0.6V. The horizontal-sweep generator is set at 0.5 msec per division. The period of the sawtooth is 2.5 msec.

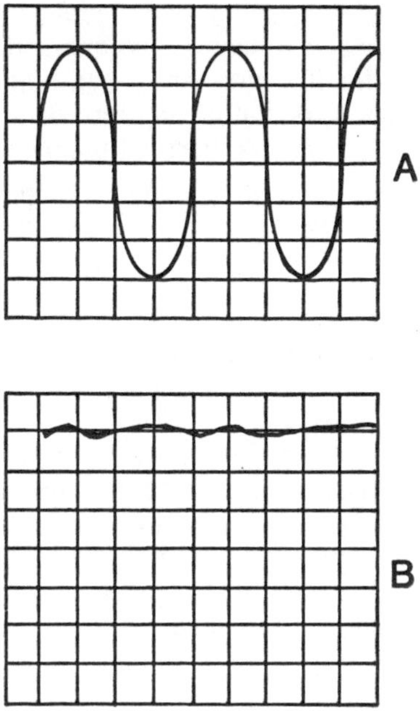

Fig. 4-19. Measurement of the peak-to-peak amplitude of an AC signal riding on a DC offset. (A) The oscilloscope is AC coupled, with a deflection sensitivity of 10 mV per division. A sine wave of 80 mV peak-to-peak amplitude is measured. (B) The oscilloscope is DC coupled, and the attenuator is set to give a deflection sensitivity of 1V per division. The 80 mV AC signal can be seen riding on a 3V positive DC offset.

one division, a multiplication of the time-per-division settings and the number of divisions involved is required. If the magnifier is employed, the final period measurement must also include the magnifier multiplier.

The period of a waveform is defined as the time required for the waveform to completely repeat itself. With a triggered oscilloscope having a calibrated time base, this measurement may be easily made. Referring to Fig. 4-18, note that the length of the sawtooth is 5 divisions. The time base setting for this display is 0.5 msec per division. The period of the waveform is 0.5 msec per division multiplied by 5 divisions, or 2.5 msec. The reciprocal of period will yield frequency, in this example, 400 Hz.

Rise-time Measurements

As previously indicated, rise time is the time required for the leading (positive-going) or trailing (negative-going) edge of a wave to pass from 10% to 90% of its full amplitude. When making rise time measurements, it is preferable to expand the trace, by time base adjustment, to spread the rise or fall of the signal over as many horizontal divisions as possible. A slope of 45° is the most desirable, although not always attainable.

The signal should also be spread over as many vertical divisions as possible. The variable vertical attenuator may be used to reduce the amplitude of the signal to an integral number of divisions (five for example). Such an adjustment will facilitate measurement of the 10% and 90% amplitude points. Use of the variable vertical attenuator may introduce some error in the rise-time measurements. This adjustment should not be used if extreme accuracy is desired or if the measurement is near the rise-time specification of the oscilloscope.

Dual-Trace Measurments

When using a dual-trace oscilloscope, the operator must decide on alternate or chopped mode and on trigger source selection (channel 1, channel 2, or mixed).

Some dual-trace oscilloscopes require the operator to select between chopped and alternate modes, but this is done automatically on other oscilloscopes. The optimum point to switch from chopped to alternate mode will vary slightly with the waveforms to be observed; however, changeover as the time base reaches 1 msec per division is a good choice.

Chopped mode at high speeds will produce a square wave on the CRT with no signal applied to the oscilloscope. This can be confusing to the operator. Triggering selection usually offers the alternatives of one of the two vertical input channels or the combination of both channels. If the two signals to be observed are time (phase) related but have a frequency difference, triggering should be selected from the lowest frequency channel. For example, Fig. 4-20 shows a dual-trace display of the input and output waveforms of a BCD counter. The upper trace is the input, and the lower is the output. Triggering is selected to be on the negative slope of the lower waveform thus insuring only one sweep for each output pulse.

If two signal channels have no time or phase relationship and may have a frequency difference, the *Y1-plus-Y2* (combination) mode may be desired.

One of the most common uses of the dual-trace oscilloscope is for comparing the output and input waveforms

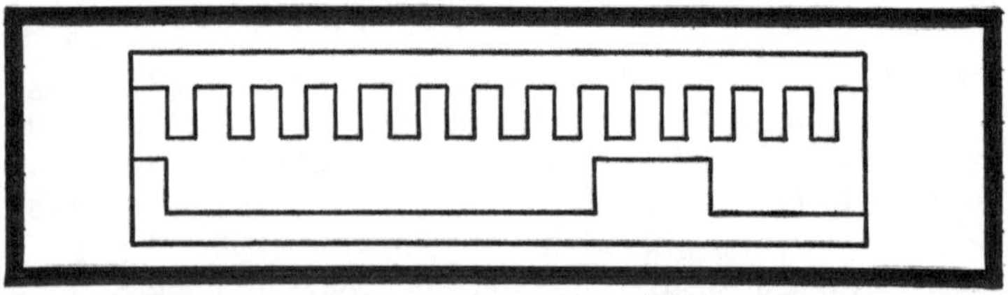

Fig. 4-20. The characteristic wave forms from a BCD counter. Triggering is on the trailing (−) edge of the lower (D-output) waveform. Always trigger on the lowest frequency signal when observing frequency ratios.

of a circuit for general changes in the shape. The POSITION controls may be used to superimpose the input and output waveforms, making differences quiet noticeable.

Another useful dual-trace measurement is the determination of time differences. Figure 4-21 shows a 300 nsec delay between the leading edges of the upper and lower pulses. A common source of error in time measurements is improper trigger source selection. The triggering must be made from the *start* signal. Utilizing the *stop* signal as a trigger source would yield a measurement of stop-to-start time, which may or may not have meaning. Think in terms of cause and effect when selecting the trigger source for such measurements.

XY Frequency Measurements

An application that is older than the oscilloscope itself is a measurement of frequency by the use of Lissajous pattern. It can be shown geometrically that plotting a sine wave of frequency *A* against a cosine wave of frequency *A* (sine wave with 90° shift) will result in a circle. This leads to a technique for frequency measurement. A reference frequency is applied to the horizontal axis of an oscilloscope (operated in the XY mode), and an unknown frequency of the same amplitude is applied to the vertical axis. When a perfect circle or something close to a perfect circle is achieved, the two frequencies are identical.

The difference between the frequencies of the X and Y signals, for small differences, may be measured by watching the circle alternately collapse and expand (rotate). Two circles will be produced each second for each cycle of difference. For example, suppose the horizontal axis is

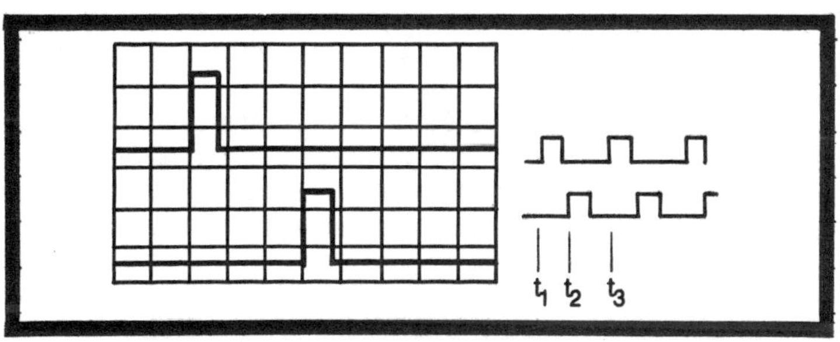

Fig. 4-21. Measurement of time separation between two pulses. The oscilloscope time base is set at 100 nsec per division. The leading edge-to-leading-edge time is, therefore, 3×100 nsec or 300 nsec. Trigger selection must be taken from the upper waveform, or measurement of the time interval t_2 to t_3 would occur rather than measurement of the time interval t_1 to t_2. Triggering selection must be influenced by the measurement to be made.

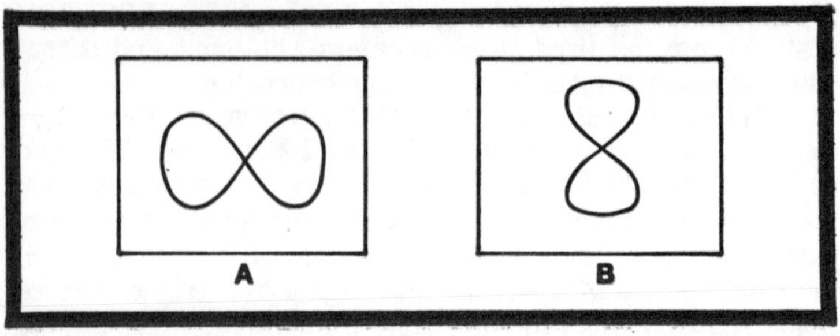

Fig. 4-22. Lissajous patterns for a 2:1 frequency relationship. (A) Vertical frequency is twice the horizontal. (B) Horizontal frequency is twice the vertical.

supplied with a 100 kHz reference frequency and the unknown frequency is applied to the vertical axis. If in 10 seconds four perfect circles are made, there is a 2-cycle-per-10-second, or 0.2 Hz, difference between the two frequencies. The measurement does not determine whether the reference frequency is higher or lower in frequency.

This technique can determine that the frequency difference is less than a certain amount. Such a technique is frequently used for setting oscillator frequencies within a certain tolerance. The upper frequency limit of this technique is dependent on the reference frequency available and the horizontal-frequency capability of the oscilloscope.

The Lissajous technique may be expanded to measure frequency ratios. Figure 4-22A shows the pattern created when the vertical frequency is twice the horizontal frequency. Figure 4-22B shows the opposite case, where the horizontal frequency is twice that of the vertical frequency. Fairly complex frequency ratios ay be measured by this technique; but as the ratios grow larger it is more difficult to maintain sufficient pattern stability long enough to determine the ratio. For most cases, ratios of 5:1 are fairly reasonable. Lissajous patterns do represent a fairly realistic form of frequency measurement. The 60 Hz power line provides a reference with frequency stability of the order of 0.1% to 0.01%.

XY Phase Measurements

Within the frequency limitations of the horizontal and vertical amplifiers, the XY oscilloscope can be used to make phase measurements. The phase measurement technique is similar to the Lissajous frequency measurement. Horizontal and vertical signal amplitudes are measured, and phase angle is computed. For phase measurements, horizontal and vertical *display* amplitudes must be identical, and horizontal and

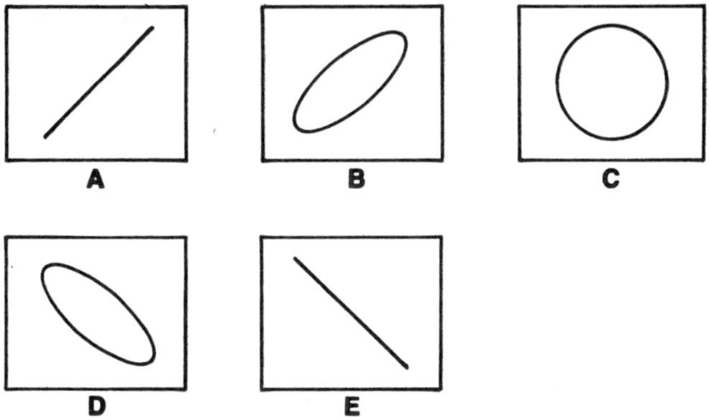

Fig. 4-23. Lissajous patterns for 45° steps from 0° to 360°—(A) 0° and 360°, (B) 45° and 315°, (C) 90° and 270°, (D) 135° and 225°, and (E) 180°. The horizontal and vertical amplitudes must be identical to produce these patterns.

vertical frequencies must be the same. Independently adjusting each channel, using the variable attenuator to set the display amplitude at an equal number of divisions, is a simple method of assuring identical amplitudes. Figure 4-23 shows the various 45° phase relationships.

The exact phase relationship may be calculated as shown in Fig. 4-24. Phase measurements made on an XY oscilloscope must include any phase difference between the X and Y channels of the oscilloscope at the measurement frequency. For comparison purposes, Fig. 4-25 shows a 1% phase shift.

APPLICATIONS

Although this section can by no means be a comprehensive discussion on all oscilloscope measurement possibilities, it reveals a number of rule-of-thumb techniques which permit

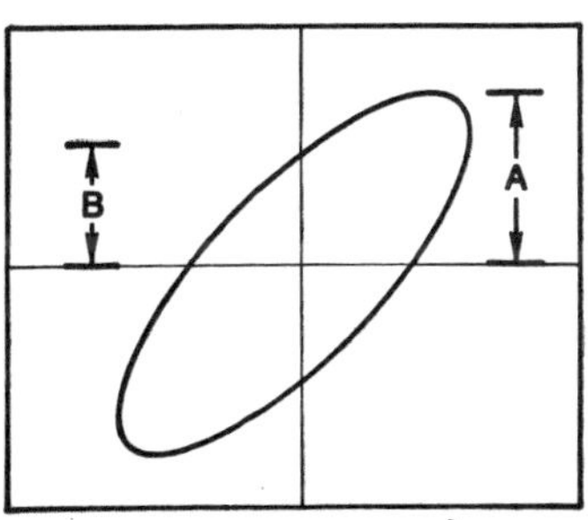

$$\theta = \text{ARC SIN } \frac{A}{B}$$

Fig. 4-24. The measurement of an arbitrary phase angle from a Lissajous pattern. The X and Y amplitudes must be equal for the relationship to hold.

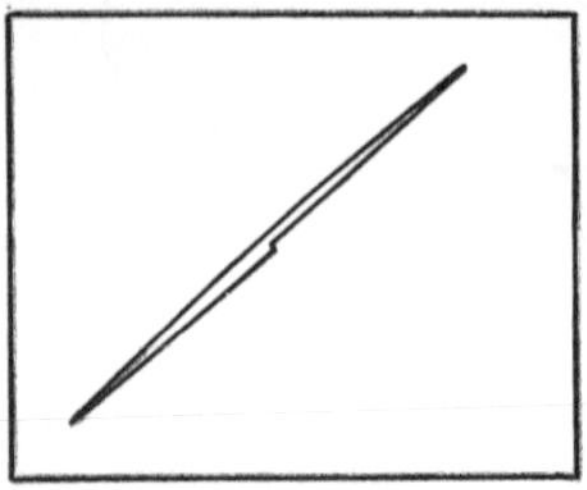

Fig. 4-25. A Lissajous pattern with 1° phase shift between the X and Y axis. This is the limit of resolution for most Lissajous phase measurements.

rapid analysis of many different circuits. These are primarily concerned with servicing and with analyzing problems of common circuits.

Amplifier Response Testing

Analysis of an audio amplifier is made by introducing a square wave to the input jack of the amplifier and observing the amplifier output waveform. Figure 4-26A shows a dual-trace comparison of the input (upper) square-wave signal and the output (lower) waveform. Lack of low-frequency response is evidenced. Figure 4-26B shows the opposite problem, lack of high-frequency response. The effects of Fig. 4-26A may also indicate too low a frequency being used to permit AC vertical coupling. Figure 4-26B may also indicate that the oscilloscope is being used near its upper bandwidth limit. Before testing an amplifier or other device for square-wave response, it is advisable, especially with a single-trace oscilloscope, to check the appearance of the square wave at the highest and lowest frequency limits to be used. If the oscilloscope itself will degrade the waveform, all

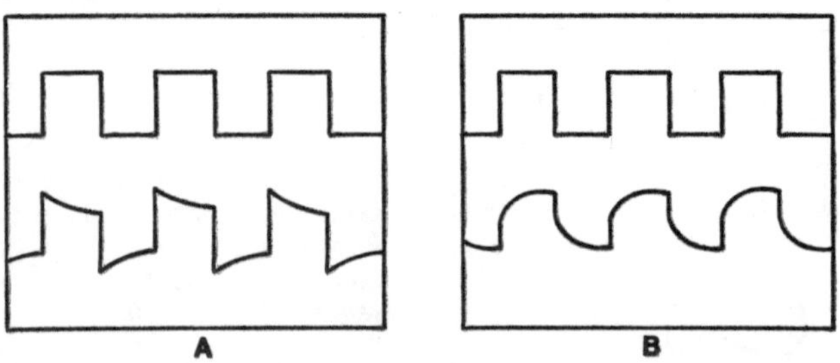

Fig. 4-26. A square-wave frequency-response test. (A) Poor low-frequency response is indicated. (B) Poor high-frequency response is indicated.

measurements must reflect the degradation by the oscilloscope.

When determining the frequency response of an amplifier in detail by performing a point-by-point plot of frequency response, a dual-trace oscilloscope may be used as the readout device. The input sine wave is monitored on one channel, and the output sine wave is monitored on the second channel. As the test frequency is increased, the input amplitude may be readjusted to a constant deflection value. The output amplitude is measured and recorded. The bandwidth of the oscilloscope must at least equal that of the amplifier being tested.

If a square wave with a fast rise time is available, the rise time of the amplifier may be readily measured. If the amplifier rise time and its 3 dB frequency are known, the high-frequency characteristics of the amplifier are defined. An amplifier showing a rise-time—bandwidth product much greater than 0.35 will have poor upper high-frequency response, while one showing substantially less than 0.35 will produce overshoot (undesirable high-frequency response).

The dual-trace oscilloscope may be used to measure phase delay through an amplifier. A sine wave is applied to the amplifier and is monitored on one channel. The output waveform is monitored on the second channel, and the time differential between the two waveforms is observed. The measured time delay, divided by the period of the input signal and multiplied by 360°, will yield the phase shift in degrees. Lack of a uniform phase shift across the amplifier bandwidth indicates a nonuniform frequency response. This will contribute to the amplifier's total distortion.

In many audio and industrial amplifiers the complementary-pair output state is very popular. Complementary amplifiers are prone to two failures; both are readily identifiable. With a sine wave applied to the input and with the output observed on an oscilloscope, complete failure of one member of the complementary pair will produce an output signal with an appearance like that of the output of a half-wave rectifier. The polarity of the missing cycle indicates the area of difficulty.

The second problem, not quite as obvious or as easy to repair, is caused by improper biasing. To operate properly, the complementary pair must switch between the positive-going output device and the negative-going output device exactly as the output waveform passes through zero. If there is a biasing failure, this does not happen. The result is called *crossover distortion.* A detailed examination of the zero-crossing portion of the waveform will show a step, indicating crossover distortion.

Partial failure of a power supply or input overdrive will result in clipping. Again, observation of the output waveform can rapidly identify either single-sided clipping, which is most frequently indicative of a power supply or biasing failure, or symmetrical clipping, which is indicative of total power supply failure or input overdrive.

Severe harmonic distortion can be detected with an oscilloscope as a noticeable deterioration of the shape of a sine wave. With a dual-trace oscilloscope, the output sine wave can be compared to the input sine wave by superimposing them. Distortions of less than 3% are very difficult to see on an oscilloscope. Below 3% one should rely on a distortion analyzer.

Analysis of Logic Circuits

The analysis of logic circuits without an oscilloscope, from either a design or servicing standpoint, can be close to futile. In fact, anyone attempting to design or service logic circuits regularly should certainly consider obtaining a dual-trace oscilloscope. Although vertical-delay lines are not an absolute must, they will add considerably to logic analysis. Logic circuits usually have timing and pulse characteristics that cannot be ignored. When a logic circuit fails, and checks of the power supply and other such basic items indicates a logic problem, the first area to check is the amplitude and timing characteristics of the driving pulses. Frequent failure points for digital equipment are the circuits which convert signals from analog levels to the levels and shapes required for the digital circuits.

With some of the slower logic families, especially MOS, pulse-width measurements may reveal a pulse too narrow for the logic family in use.

A note of caution is required here. Many common logic circuits operate at high speeds. Narrow pulses may easily be lost because a low-cost oscilloscope does not have a bandwidth to display them. Figure 4-27 shows pulses observed on a high-speed (50 MHz) oscilloscope (A) and the same pulses on a medium-speed (15 MHz) oscilloscope (B). An oscilloscope with a bandwidth of 15 MHz or greater is the recommended minimum when servicing TTL circuits.

When attempting to observe spiking, the oscilloscope must have extended bandwidth. In a TTL circuit, a 15–20 nsec spike is sufficient to trigger a flip-flop, but it would not be observed on a 10 MHz oscilloscope and would be of reduced amplitude on a 15 MHz oscilloscope.

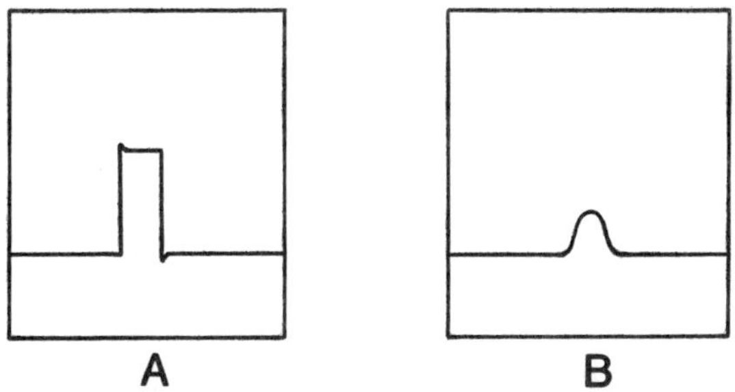

Fig. 4-27. The effect of vertical bandwidth on pulse response. (A) a 25 nsec pulse as seen on a 50 MHz oscilloscope, (B) the same pulse as seen on a 15 MHz oscilloscope.

Spiking can also occur directly after a mechanical contact closure (contact bounce). This would be observable as a number of spikes following the actual contact. However, this is very difficult to observe because of the narrowness of the spikes and the infrequent operation of the mechanical switch. A storage scope is required to observe this phenomenon.

When analyzing circuits involving counters, a frequent problem is excessive counts. There are two common causes of excessive counts. Ringing (see Fig. 4-28) will sometimes produce sufficiently large spikes to cause a second count. Oscillation, or noise, in analog interface circuits can frequently cause excessive counting. Often these result in conditions similar to those shown in Fig. 4-29. When examining ringing conditions in logic circuits, proper use of the probe and

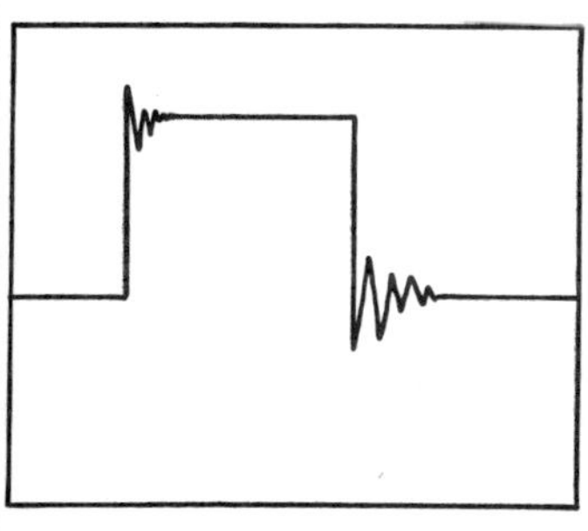

Fig. 4-28. Ringing at the leading and trailing edges of a pulse in a logic circuit. Ringing may be present within the logic circuit. Ringing may also be caused by the measurement apparatus itself. If the amplitude and shape of the ringing appear to change substantially while you run your hand up and down the test cable, suspect the measurement technique. Ringing of this sort may also change considerably with changes in placement of the ground clip.

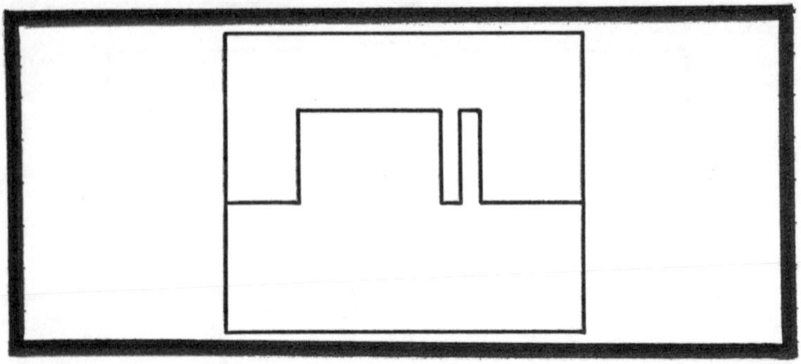

Fig. 4-29. One of the effects of extensive ringing in a logic circuit. Limitations in vertical bandwidth may reduce the amplitude of this pulse. An input–output comparison of the succeeding stage with a dual-trace oscilloscope may reveal the true effect of the pulse.

its accompanying ground clip are extremely important. A long ground clip, or one attached at some distance from the point of measurement, may cause ringing in the measurement itself. When attempting to observe logic circuits, a ×10 (low-capacitance) probe should be used. Direct measurement with a coaxial cable frequently presents an excessive capacitance to the logic circuit, causing flip-flops, for example, to become inoperative.

Figure 4-20 shows the input and output waveforms of an 8421 BCD counter. The upper trace represents the input waveform; the lower trace represents the output. To insure proper triggering, the oscilloscope is triggered on the trailing edge of the output waveform, on channel 2. The synchronization of the input waveform is then automatic. Knowledge of the clock line and the cause-and-effect relationship of the various logic circuits being examined is a great aid in establishing proper triggering.

Logic circuits often present situations demanding great writing speeds from an oscilloscope. A typical situation is the need to display a very short timespan, such as 100 nsec per division, when the triggering repetition rate is low—for example, 1 msec. To overcome this, turn the intensity up, turn the graticule illumination down, and use a viewing hood. There is a point beyond which even the finest oscilloscopes will not write because of insufficient repetition rate. Unfortunately this situation is not uncommon with logic circuits. The storage oscilloscope is a partial answer, but it lacks adequate bandwidth to observe logic circuits. Where possible, the best solution to this problem is modifying the circuit being tested to temporarily increase the basic repetition rate sufficiently to observe the desired waveforms.

Television Analysis

The complex waveform containing synchronizing and video information from a television transmission cannot be analyzed without the aid of an oscilloscope. As noted earlier, the oscilloscope is so important to television analysis that two specialized forms of the instrument have been created.

Complete vertical and horizontal presentations are shown in Fig. 4-30. The vertical presentation (A) readily shows the equalizing pulses, and the horizontal-triggering pulses show as a dim image at the synchronizing level. On the horizontal presentation (B) there is often a faint trace of a second horizontal pulse midway between the two bright horizontal pulses. This is due to effects of interlaced scanning. The horizontal presentation shows the 3.58 MHz color burst after the leading edge of the horizontal synchronizing pulse.

Signals such as in Fig. 4-30 should be observable at the output of the video detector and through the video stages of most television sets. The synchronizing signals can be easily traced through their respective circuits in case of difficulty in these areas.

When the oscilloscope is used with a sweep generator to examine the bandpass characteristics of high-frequency circuits, it is used simply as an XY display. A typical case is shown in Fig. 4-31. The displayed signal gives the amplitude-versus-frequency characteristics of a 45 MHz IF amplifier of a television receiver. In this particular case, the sweep generator and its built-in demodulator provide the vertical signal as well as the sweeping signal. Using the sweep from the generator insures that the oscilloscope sweep is always in phase with, and has the same distortions as, the sweep changing the frequency of the signal generator. An alternative method is to use the output signal from the sweep generator to synchronize the sweep oscillator of the os-

Fig. 4-30. Vertical and horizontal presentations of a color television test signal.

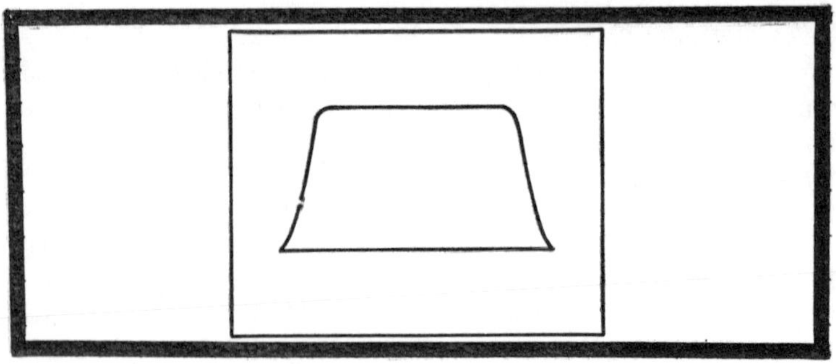

Fig. 4-31. The oscilloscope used as a display for a sweep generator–demodulator. The line across the bottom represents the sweep return.

cilloscope; however, if the sweep generator is swept by a sine wave, as is common on low-cost sweep generators, the oscilloscope must be swept by a sine wave as well, or improper frequency-versus-position relationships will exist in the display.

Figure 4-32 shows the display generated on a vector oscilloscope monitoring a color television set driven by a color bar generator. If the oscilloscope is not specifically designed as a vector monitor, an XY oscilloscope may be used. Horizontal bandwidth of at least a few megahertz is required for this application. The vector monitor may be used to adjust the amplitude of the 3.58 MHz oscillator; adjust the 3.58 MHz reactance coil, exactly phasing the 3.58 MHz oscillator with the broadcast 3.58 MHz reference; adjust the color phase detector; check demodulation angle and linearity; and adjust the chroma bandpass of the receiver.

Many solid-state TV receivers supply both chroma and luminance information to the cathode of the CRT. The signal on the vector monitor will appear fuzzy unless the luminance signal is disconnected. This can be achieved by bypassing the luminance signal with a large electrolytic capacitor at an earlier stage.

Fig. 4-32. A vectorscope presentation of the rainbow pattern produced by the red and blue of a TV with luminance signal removed.

Power Supplies

Problems with power supplies, especially regulated supplies, must be detected with an oscilloscope, not a meter. A loss or reduction in value of the filter capacitor causes excessive ripple. Ripple is easily examined using AC coupling and the high-sensitivity positions of the input attenuator. A dual-trace oscilloscope, with one channel monitoring the output of the regulated power supply and the other monitoring the input to the regulator circuitry, shows spikes on the output voltage if the input ripple dips below the regulator's capability.

A frequent problem with regulated power supplies is oscillation. Regulated power supplies are high-gain amplifiers, and loss of a frequency-compensating component, such as a bypass capacitor, can cause a power supply to oscillate. The power supply may exhibit sluggish regulation characteristics as well as a high-frequency output signal. Frequently large capacitive or inductive loads will cause a power supply to oscillate. This can even occur with bench supplies.

MEASUREMENT ERRORS

As the oscilloscope is one of the most versatile electronic instruments, it is also one of the more complicated to operate properly. Errors are frequently made, and most of them are unnecessary. Errors can be divided into two categories: those of simple operator error, and those where the display is read properly but the measurement is made improperly.

Operator errors are usually the result of an improper assumption on the part of the operator. A typical source of error is the use of an oscilloscope without a prior check to insure that the horizontal and vertical variable controls are in their calibrated positions. This problem is common in shops and laboratories where the oscilloscopes are shared by a number of users. In this situation a moment spent checking all settings on the oscilloscope, especially those not used frequently, is well worthwhile. Other errors of a similar nature include improper trigger selection in timing measurements, failure to include probe attenuation in amplitude measurements, and errors made when rapidly converting time to frequency.

Measurement errors often result from lack of understanding of the equipment specifications. For example, the error resulting from using an oscilloscope of too low a bandwidth to analyze a logic pulse may cause the operator to miss a narrow pulse causing all the problems. Errors of this type are common on both horizontal and vertical measurements when the operator forgets the accuracy specification and assumes he has made an absolute measurement.

When connecting the probe or test cable to the circuit, the effect on the circuit of the input impedance of the oscilloscope and the test lead, or probe, must be remembered. *The simple coaxial cable has a capacitance of 30 pF per foot.* With the customary 3 ft cable and 30–40 pF input capacitance of the oscilloscope, there is a total of 120–140 pF loading placed on the circuit. With certain low-impedance circuits, severe circuit loading occurs as the test frequency is increased.

When connecting the probe or test cable, the operator must be certain the ground clip is connected to a point which is grounded for the circuit under test. A frequency problem with products having both logic and analog circuits is high-level logic currents in certain portions of the ground circuits. If this is the case, and a measurement is being made in the analog circuit, the proper ground point is one which does not contain current from the logic system. For this reason the common technique of connecting the ground clip to the chassis is not wise.

The problem is even more serious as oscilloscopes become available with high sensitivities. For example, voltage drops in a foil due to return currents from a logic chip are generally not evident on an oscilloscope of 10 mV sensitivity, but are a serious problem on an oscilloscope with 1 mV sensitivity.

The compensated ×10 oscilloscope probe can introduce considerable error to a measurement involving high-frequency characteristics if the user fails to adjust the probe to the input capacitance of the oscilloscope. This procedure must be carried out even if the probe is moved from one channel of a dual-trace oscilloscope to another.

When possible, probes should be purchased for a particular oscilloscope, matched to a particular input channel, identified as such, and never used anywhere else. Unfortunately this is not always possible for economic reasons. In such cases the user must be sure to check the compensation of the probe.

CALIBRATION

An oscilloscope, like any other good instrument, should be recalibrated on a regular schedule. This is usually every 6 months to a year, depending on the need for ultimate accuracy and the history of the particular oscilloscope. Even if a full recalibration is not performed, a check of the more important specifications should be carried out periodically. The time base accuracy can usually be checked at one point for a rough test, and the 60 Hz line will usually be accurate enough. If possible, a check should be made at 1 kHz, using a square wave. The vertical accuracy can be checked in the DC-coupled mode using a known DC source. High-frequency compensation problems can be discovered by using a square-wave source of 100 kHz to 1 MHz with a known waveshape.

Most high-sensitivity DC-coupled oscilloscopes have front-panel provisions for DC balance adjustments. This permits the user to balance out any DC offsets generated in the vertical amplifier when the variable vertical attenuator is used. To adjust the DC balance, proceed as follows:

1. Center the trace with the VERTICAL POSITION control.
2. Change the variable vertical attenuator from its CAL position to the extreme opposite end of its rotation.
3. Return the trace to the center, using the VERTICAL BALANCE control.
4. Return the variable vertical attenuator to the CAL position.
5. Center the trace with the POSITION control.
6. Repeat steps 1 through 5 until there is no movement of the trace as the variable vertical attenuator is adjusted.

Frequently an oscilloscope with a DC BALANCE control will also have an internal step BALANCE control to remove DC offsets caused by changes in the step attenuator. Adjustment of the step BALANCE control, if it is to be done, should be done before the variable balancing is done. The procedure is identical to that of the variable vertical-balance adjustment, except the change in vertical sensitivity is done by moving the attenuator between the most sensitive and the next most sensitive positions.

SUMMARY

The oscilloscope has long been a valuable tool in the research and development laboratory. Unfortunately, the price of this instrument has prevented extremely useful, sophisticated versions of this instrument from making their appearance in the consumer products service shop. Modern technology is rapidly changing this situation. The high-performance oscilloscope is falling in price and consumer electronic products are becoming much more complex. The combination of these two factors leads to the more rapid acquisition of fairly complex oscilloscopes by the service shop.

I hope that some of the material presented in this chapter enables the proper selection of such an oscilloscope. The purchase of a good oscilloscope will not alone insure its proper use, even that it will be used at all; but the fundamentals of oscilloscope operation and use as explained here should make it easier for the technician to lay aside his VTVM and use the oscilloscope with confidence.

Chapter 5
Curve Tracers

The curve tracer is an instrument which has been in use for some time, although usually in the research laboratory and not the service shop. There are two reasons for this. First, there was no real need for analysis of active circuit components in the service business so long as such devices were few and simple in nature (e.g., vacuum tubes, diodes, and simple transistors). Second, curve tracers have been expensive and so were among the last instruments to be acquired by a service facility.

Time and instruments have changed. Electronic products today, even those for consumer use, no longer contain a few active devices whose characteristics are well known. Instead they have many different devices of disparate characteristics to be replaced, substituted for, or matched in order to properly service the product. Solid-state devices have caused these servicing problems because they come in many different types, and most important, the characteristics of devices with identical part numbers will vary by factors of two to one or more. This often means servicing by substitution, not exact replacement. The curve tracer has thus become an important tool.

The curve tracer will help to solve many servicing problems. If a device is to be replaced and there is a question of the proper replacement, the curve tracer can be used to determine the exact characteristics of the new device. If a device cannot be identified as to type, the curve tracer will reveal the unique identification curves of the component and hence its suitability.

Because of increased demand and improved electronic technology, prices on curve tracers have dropped drastically—from many thousands of dollars to less than a thousand dollars. Adapters are available for less than a thousand dollars. The problem is no longer the cost of acquisition of a curve tracer or adapter; it is now learning how to use the instrument to provide maximum return on the investment.

FUNDAMENTALS

The curve tracer draws a graph which is a picture of how the output current of a device varies with changes of input voltage (or current) and output voltage. The user controls the input signal and the output voltage, the curve tracer draws the graph. This graph has output current as the vertical axis and output voltage as the horizontal axis. A new curve on the graph is drawn for each input variation desired. The result is a family of curves, such as those shown in Fig. 5-1 for a simple NPN transistor.

Figure 5-2 is a simplified block diagram of a curve tracer. The vertical amplifier, horizontal amplifier, and CRT comprise a simple low-bandwidth oscilloscope. The curve-tracing sections are the sweep supply and the step generator. In this particular case an NPN transistor is shown as the device under test. Sweep voltage is used to provide increasing voltage across the transistor (between the C and E terminals) as well as to deflect the oscilloscope beam from left to right. Generally, this voltage will be full-wave-rectified signals derived from the line via a transformer. A switch sets various voltage ranges, and a variable control selects the value of sweep voltage within this range. A sweep voltage is applied to the transistor through limiting resistor R_{il} to protect the device. Note that the horizontal deflection signal for the oscilloscope is connected so the beam deflection represents the actual voltage being impressed across the device. After each complete cycle of the sweep voltage, the current (or voltage) of the step generator is advanced to the next higher step. Resistor R_{im} in series with the sweep supply converts the sweep current to a voltage. This voltage is applied to the vertical amplifier of the oscilloscope to provide vertical deflection that is proportional to sweep current.

During the first sweep an increasing voltage is applied to the collector of the transistor; but there is no base current, as

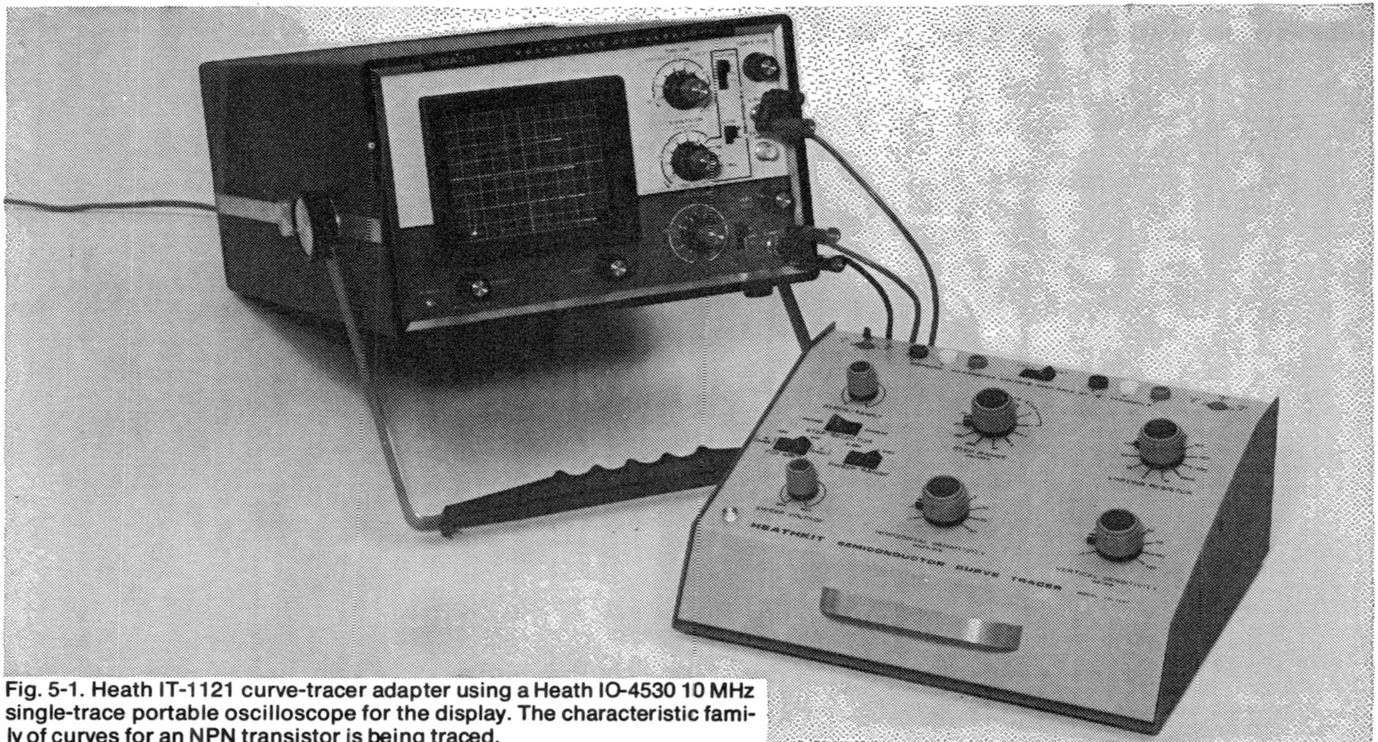

Fig. 5-1. Heath IT-1121 curve-tracer adapter using a Heath IO-4530 10 MHz single-trace portable oscilloscope for the display. The characteristic family of curves for an NPN transistor is being traced.

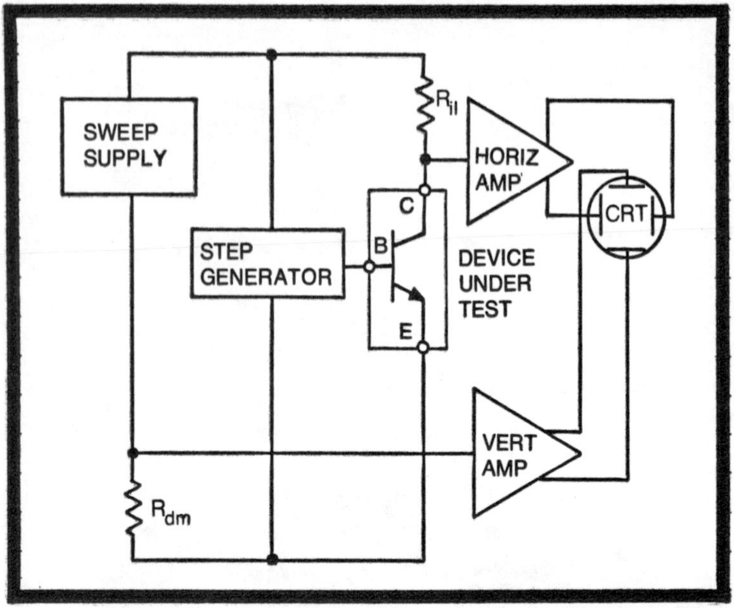

Fig. 5-2. A simplified block diagram of a curve tracer. The sweep supply and step generator develop the special waveforms used to exercise the device under test. The horizontal and vertical amplifiers as well as the CRT comprise the display.

the step generator is at the zero step. As the sweep voltage is increased to its peak value and then returned to zero, the spot on the CRT, which was initially positioned at the lower left corner of the tube face, moves across the face to the lower right corner and then returns to the lower left corner. Assuming that the breakdown voltages of the transistor are not exceeded and that there are no large leakage currents, there will be no collector current in the transistor during this first sweep; so there will be no vertical deflection, and a straight line will be drawn (A of Fig. 5-3).

When the first sweep is complete, the step generator is advanced from the zero step to a level of current selected as the step value. As the sweep moves the spot across the CRT and applies voltage to the collector of the transistor under test, collector current will commence (because there is base current). This collector current causes a voltage drop on the current-sensing resistor R_{im}, which in turn causes a vertical deflection of the oscilloscope trace during this second sweep (B of Fig. 5-3). Note that under very low collector voltages (to the left of the knee) the current through the transistor is limited by the *on* resistance of the device and the applied collector voltage. After the collector voltage has increased sufficiently, the current drawn is dependent upon the beta of the device and is essentially constant.

Succeeding sections of Fig. 5-3 show how each individual curve in the family of curves is generated as the step generator increases its value by one current unit after each sweep. Because each of these curves is being generated in 1/120 sec, the eye cannot see them being individually created. The result is the composite picture shown in D of Fig. 5-3.

Generators

The sweep generator, current-limiting resistors, and step generator values may be selected over a wide range to accomodate a variety of semiconductor devices. The step and sweep generaors both have polarity reversal capability so that NPN and PNP devices may be examined. The step generator may be changed from current steps to voltage steps to permit testing of field-effect devices such as FETs and tubes.

Adapters

Curve tracers come in two forms—as complete units or as adapters. The adapter and a simple low-cost oscilloscope make a complete curve tracer. The primary advantage of the adapter is low cost. The oscilloscope is not committed to the curve tracer when the curve tracer is not being used. For the small shop, experimenter, or infrequent user, this often represents the most economical way to obtain curve-tracing capability. The disadvantages of the adapter are the required reassembling of the curve-tracer setup each time it is to be used and its lack of portability. The question of curve tracer versus adapter is primarily one of economics.

SPECIFICATIONS

To properly select and use a curve tracer, the specifications must be understood. Specifications define the limits of measurement that may be made with a certain curve tracer, such as voltage ranges, power limits, and current resolution. Before purchasing the instrument, its specifications should be carefully compared with those of the others available and, most important, with the user's anticipated needs. A good way to do this is to make a chart listing specifications and costs along one edge and manufacturers along the other, filling it with data for the various models.

Sweep (Collector) Supply

The two primary characteristics of the sweep supply are its voltage ranges and current capability. On the more

A

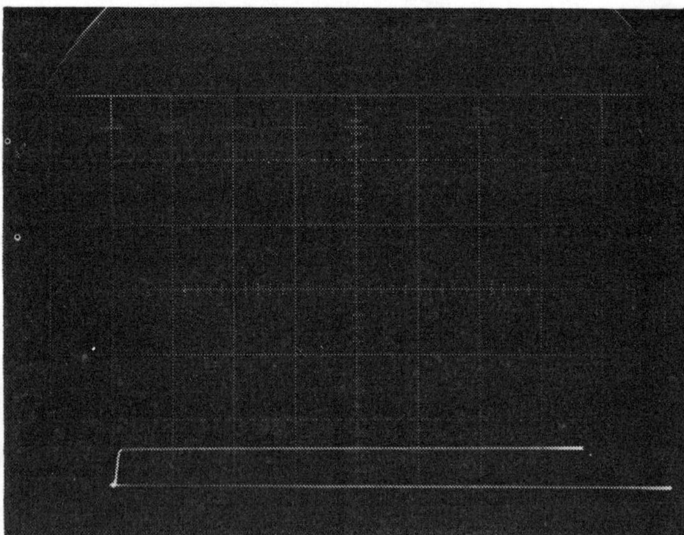

B

Fig. 5-3. The sequential development of a family of curves. The curve-tracer settings are: step current, 0.02 mA/step; collector series limiting resistance, 1000 ohms; horizontal sensitivity, 2V/CM; and vertical sensitivity,

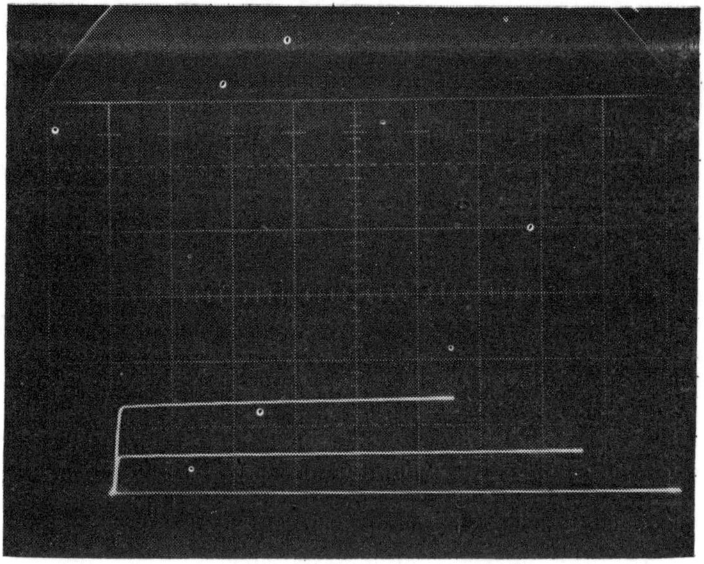

C

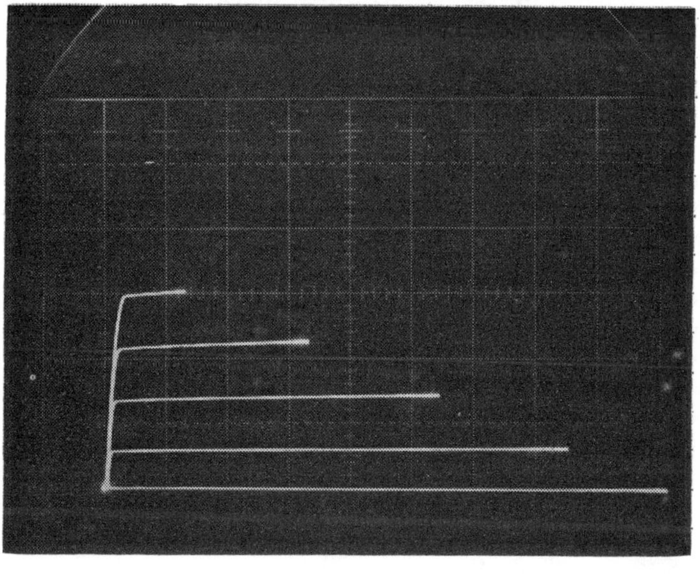

D

5 mA/cm. (A) Base current is at zero step. (B) Base current is at 0.02 mA step. (C) Base current is at 0.04 mA step. (D) All steps shown: 0, 0.02, 0.04, 0.06, and 0.08 mA.

complete (sophisticated) curve tracers there will be a number of voltage ranges, each with a maximum current level; on the less complex instruments there will usually be at least two ranges of voltage. One of the ranges should give reasonable resolution for small-signal and medium-power devices; the other range should give sufficient voltage to make breakdown tests. Typical minimum ranges might be 0V–40V at 1A and 0V–200V at 2A. More sophisticated tracers will give ranges to 1000V or 1500V. Some curve tracers provide a DC mode of operation, in which a sweep voltage of filtered DC may be obtained. The DC mode allows special low-current measurements for situations in which the sweeping action of the full-wave-rectified signal is not desirable. To measure NPN and N-channel as well as PNP and P-channel devices, the sweep voltage must be able to provide positive voltage sweeps as well as negative ones.

Sweep-Limiting Resistors

To limit current through the device under test to a reasonable maximum for the particular device, a set of series-limiting resistors is provided. These will usually range from 0 ohms to some upper value which is a function of the highest sweep voltage that may be obtained.

Horizontal Sensitivity

Horizontal sensitivity indicates the amount of sweep voltage (applied to the device under test) required to cause one division of horizontal deflection on the CRT of the curve tracer. The range of sensitivities required is bounded on the upper end by the maxmum sweep voltage, so that a curve tracer with a 0V–200V sweep and an 8- by 10-division screen would need a minimum sensitivity of 20V per division. The maximum sensitivity should allow a full horizontal presentation of no more than a volt, giving a lower boundary of 0.1V per division. Both horizontal and vertical sensitivities are usually adjustable in a 1–2–5 sequence to yield the best resolution at all settings. Sequences of 1–10–100 usually will not provide sufficient resolution for signals in the 3-to-7-unit range.

Vertical Sensitivity

Vertical sensitivity specifies the amount of current that will be required to cause one division of deflection on the vertical axis. Again, the upper boundary of this sensitivity is governed by the amount of current that the sweep generator can supply. If the curve tracer is able to supply 1A of current and the oscilloscope screen is 8 × 10 divisions, then the minimum vertical sensitivity must be at least 125 mA/cm (1A/8 cm). However, as noted before, a 1–2–5 sequence is usual on the vertical axis, and therefore a minimum sensitivity of 200 mA per division would be used. The maximum sensitivity that is to be had with a curve tracer should give full-screen deflection with a few milliamperes. The typical values run from 500 μA per division to 1 μA per division. Most work done with small-signal discrete devices can be done with a 0.5 mA per division (500 μA per division) sensitivity; however, design and evaluation work on integrated circuits will require sensitivities in the 1 μA region.

The user should note that maximum sensitivities of the curve tracer can be improved if an adapter is being used and the oscilloscope selected for the display has a deflection sensitivity superior to that required by the adapter. For example, suppose a curve tracer adapter requires an oscilloscope with a 100 mV/cm vertical sensitivity. If this adapter has a maximum vertical sensitivity of 500 μA/cm, this may be improved to 50 μA/cm by setting the oscilloscope deflection to 10 mV/cm.

Step Generator

Step generators in curve tracers are usually abe to operate in one of four modes: (1) positive current steps (2) negative current steps, (3) positive voltage steps, and (4) negative voltage steps. In the more elaborate curve trcers, the zero step voltage or current may be offset from zero volts or amps by some positive or negative DC value. Step ranges depend upon the type of devices that the user expects he will be measuring; 2 μA to 10 mA per step in a 1–2–5 sequence would be a typical spread of steps to give fair measurement capability. Voltage steps tend to be less numerous than current steps, as most measurements are confined to a smaller dynamic range. A typical range of voltage steps on a low-cost unit is 50 mV to 1V per division, again in a 1–2–5 sequence.

Associated with the step generator there will be a variable control permitting the user to select the number of steps per family. This selection will usually range from zero to a maximum of six or nine. During normal use the control will be kept at a middle to upper range setting. For detailed analysis the user may need to adjust the number of steps.

Accuracy

Accuracies found in curve tracers will generally be similar to those expected in oscilloscopes. Accuracies of ±3%

to ±5% are typical of the vertical and horizontal amplifiers and the step generator. There will usually be some additional degradation of the accuracies when a horizontal or vertical magnifier is used. Accuracy of curve tracer measurements is not the most important factor; relative measurements and the shapes of curves are far more important.

FEATURES

Additional features which may be available with a curve tracer or adapter will make certain measurements easier or more difficult. Features that simplify measurements are:

- *Dual sockets or test positions with an A–B switch.* This feature allows the user to test two similar devices for difference characteristics or to connect a device to the curve tracer for testing and then disconnect it quickly. This feature is frequently used when matching semiconductor devices.
- *External connections to the E, B, and C terminals.* This will allow the connection of clip leads to the curve tracer, permitting in-circuit measurements of devices or even large sections of the circuit itself.
- *Special clips that permit the quick connection of devices, especially diodes, for large-quantity testing.*
- *Magnifiers.* Magnifiers, either horizontal or vertical, allow the user to expand the trace (usually by a factor of 5 or 10) and examine a particular area in more detail
- *Safety cover.* On certain curve tracers the device that is being tested is placed in a socket contained in a small box whose cover must be closed prior to energizing the sweep voltage. This is done as a safety feature, as the sweep voltages can become quite high. Care must be observed when testing semiconductors on a curve tracer, as very high voltages may be present on the device.

CURVE-TRACER ADAPTERS

If a curve-tracer adapter is to be used, an oscilloscope should be found that can be dedicated to the curve-tracing application. Thus an old scope that has been retired due to weight, lack of bandwidth, or some other reason may regain its usefulness. The scope to be used needs a bandwidth of 20 kHz or more and should be DC coupled in both the vertical and horizontal channels, although AC-coupled oscilloscopes can be used with most measurements. Generally, curve-tracer adapters have minimal vertical- and horizontal-sensitivity requirements, and these requirements should be carefully noted before the adapter is obtained to be sure that the oscilloscope intended for the job will suffice. Additional sensitivity in the oscilloscope will permit magnification, giving greater resolution in measurements.

DEVICE AND CIRCUIT TESTING

There are basic operating connections for different classes of devices, as well as characteristic curves which should be familiar to the user. These basics may then be extended to similar devices or circuits as required. To simplify the setup procedure for common devices, a setup table has been provided (Table 5-1). This table should be referred to for determining connection, polarity, types of steps, and so on. Succeeding sections of this chapter are to familiarize the user with common semiconductor devices.

DIODES

Although the semiconductor curve tracer is designed with sufficient complexity to test 3-terminal devices (transistors,

Table 5-1. Setup of Curve Tracer For Various Semiconductor Devices.

Device \ Setup	Terminals C B E	Sweep	Step Generator	Initial Spot Position	Remarks
NPN	C B E	+	+ Current	Lower Left	
PNP	C B E	−	− Current	Upper Right	
N-Channel JFET	D G S	+	− Volts	Lower Left	
P-Channel JFET	D G S	−	+ Volts	Upper Right	
N-Channel MOSFET (Depletion)	D G S	+	− Volts	Lower Left	
P-Channel MOSFET (Depletion)	D G S	−	+ Volts	Upper Right	
N-Channel MOSFET (Enhancement)	D G S	+	+ Volts	Lower Left	
P-Channel MOSFET (Enhancement)	D G S	−	− Volts	Upper Right	
Forward Diode	A NC C	+	None	Lower Left	Small-signal diode rectifier
Reverse Diode	A NC C	−	None	Upper Right	Zener diode or checking reverse break
SCR Forward	A G C	+	Current	Lower Left	One half of triac
SCR Reverse	A G C	+	Current	Upper Right	One half of triac

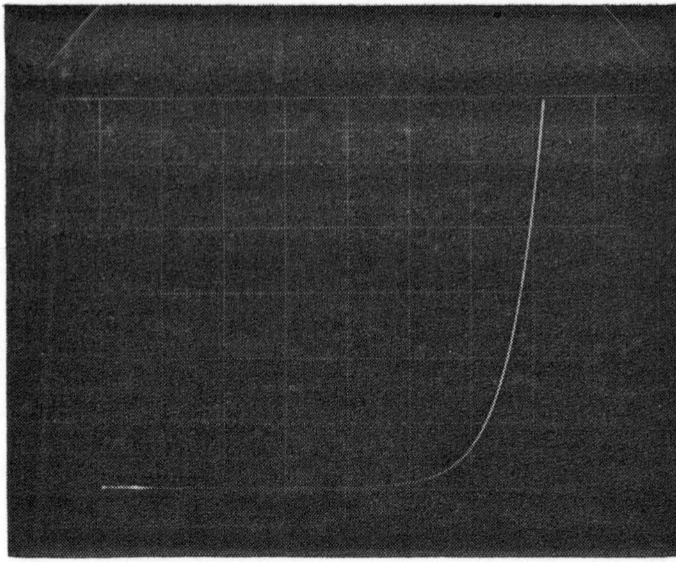

Fig. 5-4. The forward characteristics of the 1N4149 silicon small-signal diode. The curve-tracer settings are: anode series limiting resistance, 1000 ohms; horizontal sensitivity, 0.1V/cm; vertical sensitivity, 1 mA/cm. The step generator is not used in a 2-terminal measurement.

tubes, etc.), it is equally capable of testing 2-terminal devices (diodes, resistors, etc.). If a positive voltage is applied to the anode of a diode, a current flows through the diode. If a negative voltage is applied to the anode of a diode, no current will flow.

In Fig. 5-4 the forward conduction characteristics of a 1N4149 small-signal silicon diode are shown, along with the curve-tracer settings required. The diode is connected between the C (anode goes here) and the E (cathode goes here) terminals of the curve tracer. As with all 2-terminal devices, the step generator is not required. The curve on the CRT is a current-versus-voltage plot. The sweep voltage provides the variable (voltage), and the current resistor monitors variations in sweep current as a function of the sweep voltage.

On the diode curve we note that not much, if any, current flows until approximately 0.7V is reached. At that voltage the diode begins to conduct, and thereafter the characteristic becomes a straight line whose slope is the resistance of the diode in conduction. The dynamic resistance of the diode during conduction may be calculated by selecting a portion of the curve that is nearly linear, determining a change in voltage ΔV and the change in current ΔI caused by this change in voltage. The dynamic resistance of the diode at that point is then determined by

$$R_d = \Delta V/\Delta I = \frac{0.04V}{0.003A} = 13 \text{ ohms}$$

It is interesting to note that at the knee (the point where the diode begins to conduct) the diode is definitely not linear. This nonlinearity is often used for special functions, such as logarithmic amplifiers.

Germanium diodes differ from their silicon cousins primarily in the forward voltage at which they begin to conduct. As can be seen from Fig. 5-5, the *on* characteristics of the germanium diode (1N191) differ slightly from those of the silicon diode.

The curve tracer can examine curves of a reverse-biased diode also. When the diode is reverse-biased, theoretically, no current will be drawn; however, as is shown in Fig. 5-6, there is current drawn after the reverse breakdown voltage of the diode is reached. When the unknown diode is being used as a replacement part in a circuit, the reverse breakdown voltage

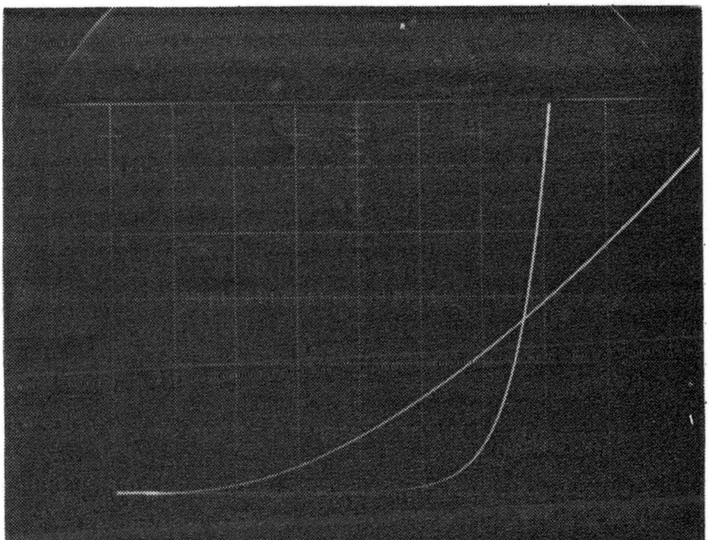

Fig. 5-5. The forward characteristics of the 1N191 germanium small-signal diode superimposed on the characteristics of a 1N4149 silicon diode. The curve-tracer settings are: anode series limiting resistance, 1000 ohms; horizontal sensitivity, 0.1V/cm; vertical sensitivity, 1 mA/cm. The step generator is not used. Note the difference in the slope and break point of the germanium diode compared to those of the silicon diode.

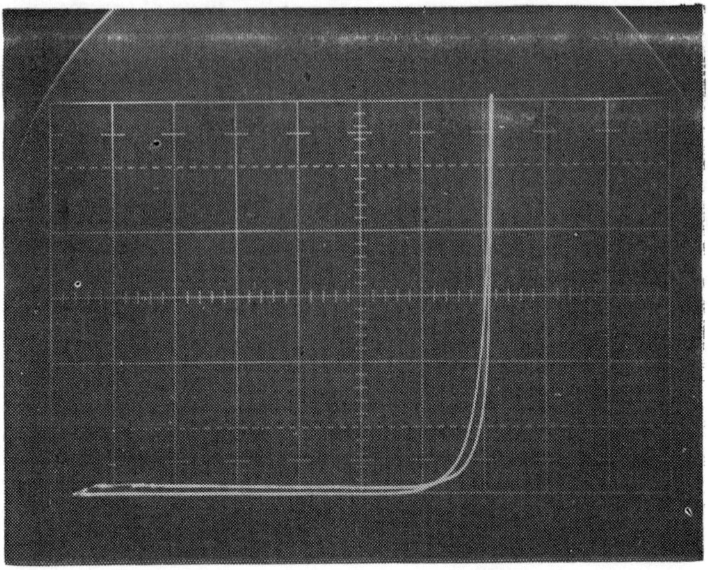

Fig. 5-6. The reverse characteristics of the 1N4149 silicon small-signal diode. The curve-tracer settings are: cathode series limiting resistance, 10K; horizontal sensitivity, 20V/cm; vertical sensitivity, 0.5 mA/cm. The step generator is not used. Note that some looping is caused by internal heating of the diode during sweep and by stray capacitances.

should be measured by using this technique. Once the diode's reverse breakdown voltage has been ascertained, the circuit parameters will determine whether or not it may be used as a replacement part. Some diodes tend to show a fairly large reverse current even before they reach reverse breakdown. This leakage current can make the diode unsuitable for a replacement part. Generally, with small-signal diodes the leakage currents will be unnoticeable on most curve tracers.

Zeners

The zener is simply a diode whose reverse breakdown characteristics are known and controlled. Figure 5-7 shows the characteristics of a 1N750 diode. Poor regulation in a zener-regulated circuit can be caused by too-high dynamic resistance when the zener is conducting. The dynamic resistance of the zener can be computed in exactly the same manner as the forward dynamic resistance of a diode. Referring to the manufacturer's published data on the diode, we can ascertain at what current the zener voltage is specified and over what current range the dynamic impedance of the zener is specified. When measurements are made in circuit,

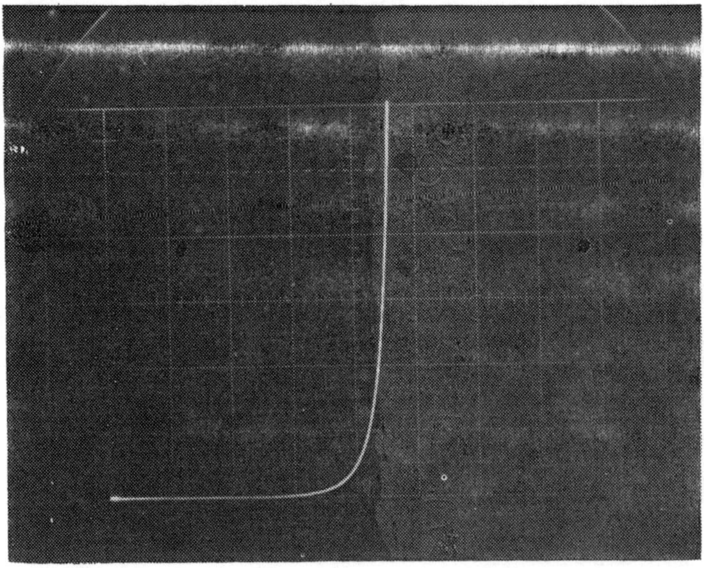

Fig. 5-7. The reverse characteristics of the 1N709A silicon zener. The diode is rated at 6.2V ±5% at a zener current of 25 mA. The curve-tracer settings are: cathode series limiting resistance, 5000 ohms; horizontal sensitivity, 1V/cm; vertical sensitivity, 5 mA/cm. The step generator is not used with a 2-terminal measurement. Note that the zener curve passes through 6.4V at the horizontal line representing 25 mA.

the dynamic resistance of the diode will be increased by the resistance of other circuit elements that are in series in the measurement circuit.

Tunnel Diode

The curve tracer is ideal for determining the characteristics of the tunnel diode. As shown in Fig. 5-8, the characteristics of the tunnel diode are unique, and one can readily identify the device by them. When using the tunnel diode, the limiting resistor may have to be adjusted to prevent oscillations in the particular tunnel diode being used. Tunnel diodes possess a negative-resistance region, wherein the current *decreases* as the voltage *increases*. Given the proper conditions, the negative resistance will sustain oscillation.

When the display has been properly adjusted and there appears to be no oscillation (identified by looping or scrolls on the trace), some of the frequently specified characteristics of a tunnel diode may be readily measured. These are: the peak voltage V_p at the start of the tunneling region, the valley voltage V_v at the end of the tunneling region, the peak current I_p at the start of the tunneling region, and the valley current

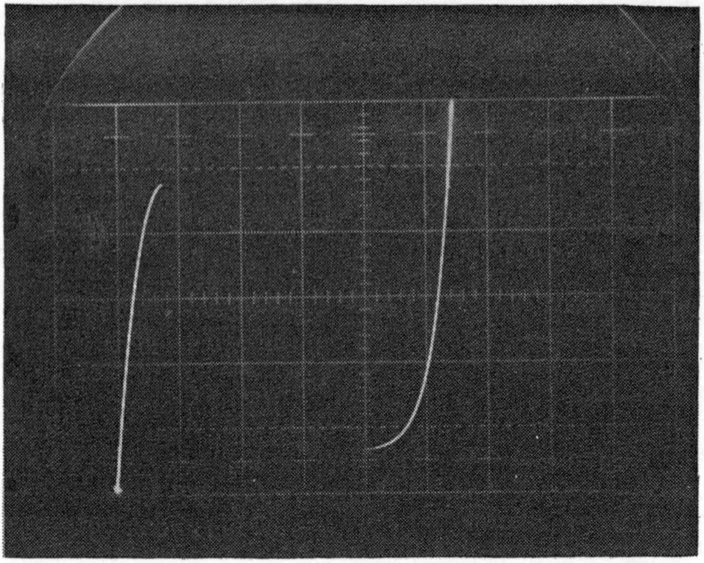

Fig. 5-8. The forward characteristics of a GE TD-716 tunnel diode. The curve tracer settings are: anode series limiting resistance, 5000 ohms; horizontal sensitivity, 0.1V/cm; vertical sensitivity, 1 mA/cm. Note the area of trace between the peak current and the valley current is almost impossible to see. If this area of the display contains scrolls, readjust the value of series limiting resistance and the sweep voltage to obtain a trace without scrolling.

I_V at the end of the current-tunneling region. Once these parameters are known, the average value of negative resistance can be computed as $(V_V - V_P)/(I_P - I_V)$. When tunnel diode curves are displayed on the curve tracer, the region between the peak and valley voltages—that is, the region of negative resistance—is frequently quite dim.

Diac

The diac, or trigger diode, which is frequently used in series with the gate of a triac or SCR to fire the device, has a characteristic curve as shown in Fig. 5-9. Note that for the diac, which is a bidirectional device, two curves are shown, as the diac has similar characteristics in the forward-bias and reverse-bias directions. The forward blocking voltage indicated on the display is that voltage where the horizontal line stops and a fine trace swings back to the high-current point, showing low forward voltage drop. As the sweep voltage is decreased, the current through the diac is reduced until it finally stops abruptly. This point gives a measurement of the holding current.

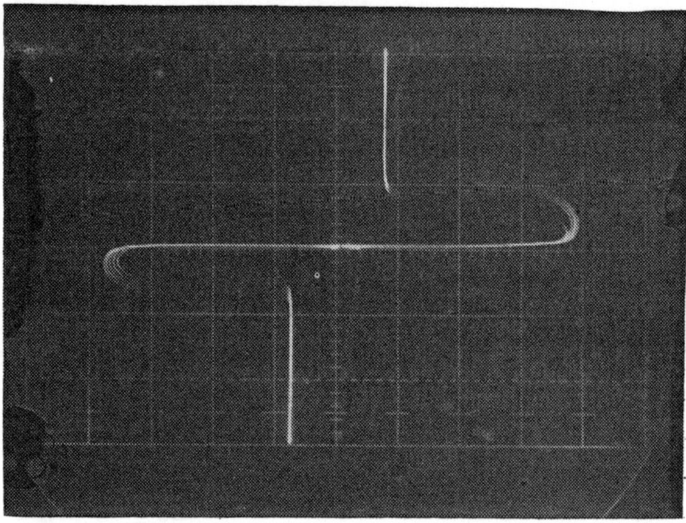

Fig. 5-9. (A) The forward and reverse characteristics of a diac or trigger diode. Both forward and reverse curves are obtained on one photograph by double exposure. Curve-tracer settings for reverse measurements are the same as for the forward measurements, except the polarity of measurement has been reversed. The settings are: series limiting resistance, 5000 ohms; horizontal sensitivity, 20V/cm; vertical sensitivity 2.0 mA/cm. Note the match between the forward and reverse breakover voltages.

BIPOLAR TRANSISTOR

A large number of characteristics of the NPN and PNP transistor may be measured with the curve tracer. All measurements applicable to NPN devices may also be applied to PNP devices; the only difference is in the initial setup of the curve tracer. To measure NPN devices, the collector or sweep voltage must be positive, and often this is shown on the curve tracer as the NPN position. The base or stepping generator must also be in the positive-current or NPN position. (Note that at this time we use a current generator, as a transistor is a current-operated device). Transistors can be very easily damaged if the voltage output of the stepping generator is used. If an unknown device is found and a curve tracer is used to ascertain the characteristics of the device, including the determination of PNP or NPN, the beam may be initially placed in the center of the curve-tracer display. Very moderate levels of sweep voltage and step current are applied first, with a midrange setting of the STEPS PER FAMILY control, first in the NPN direction and then in the PNP direction. One test will yield a family of curves, the other a straight line. The test yielding the family of curves identifies the device as NPN or PNP. Caution must be used, however, to keep both the base current and collector sweep voltage well within the reverse breakdown ratings of the transistor. If the incorrect settings are first used and the device is forced beyond the breakdown characteristics of its junction, it could be damaged.

Beta

One parameter of a transistor frequently needing measurement because of its wide variation is DC gain or beta, abbreviated h_{FE}. To a reasonable approximation, the beta of the transistor is the ratio of the collector current to the base current.

As the curve tracer gives us a family of curves, each curve represents the collector characteristics for a different base current. The DC beta may be readily determined at various collector current and voltage levels. The curves in Fig. 5-10 show that the spacing between curves caused by the various base steps is not uniform, indicating that the beta of the transistor is not uniform for large variations in collector

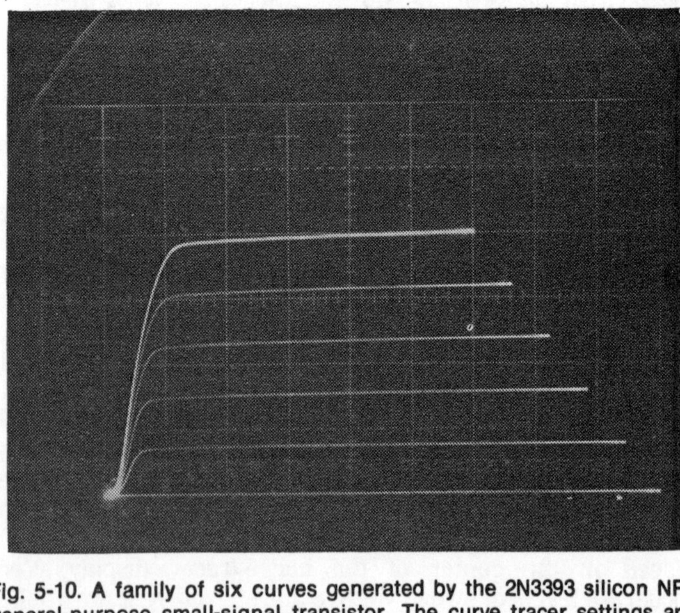

Fig. 5-10. A family of six curves generated by the 2N3393 silicon NPN general-purpose small-signal transistor. The curve tracer settings are: step generator, 0.005 mA/step; collector series limiting resistance, 100 ohms; horizontal sensitivity, 2V/cm; vertical sensitivity, 1 mA/cm. At a collector voltage of 8V the current change indicated by curves 3 and 4 is 0.8 mA. As this was caused by a step of 0.005 mA of base current, the beta of this device at this particular operating point is 0.8/0.005 = 160.

current from one step to another, caused by the change in base current from one step to another. Beta measured in this manner is often referred to as the AC beta, or h_{FE} of the device. This is a superior method of checking a transistor because the test is made near the operating current and voltage for the device. It should be noted, however, that this is not the high-frequency beta of the device, as this is only being checked at 120 Hz and is by no means representative of the beta of the device at a few kilohertz or higher.

Breakdown Voltages

Once it has been determined that a replacment transistor has sufficient beta, it is necessary to determine if it has sufficient voltage beakdown characteristics to operate successfully in the particular circuit. The characteristics in question are the collector-to-emitter voltage breakdown (V_{CE}) and the collector-to-base voltage breakdown (V_{CB}). The curves in Fig. 5-11 show the area of *collector breakdown*, defined as the area where the collector current becomes independent of the base current and begins to rise very sharply.

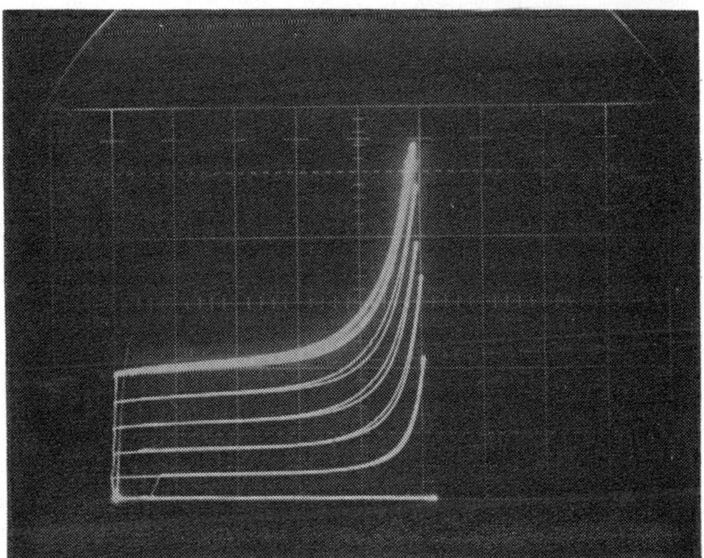

Fig. 5-11. A family of curves for the 2N3393, showing collector breakdown. Collector current becomes independent of the base current as the collector voltage exceeds the 30V–40V region. The curve-tracer settings are: step generator, 0.01 mA/step; collector series limiting resistance, 200 ohms; horizontal sensitivity, 10V/cm; vertical sensitivity, 5 mA/cm.

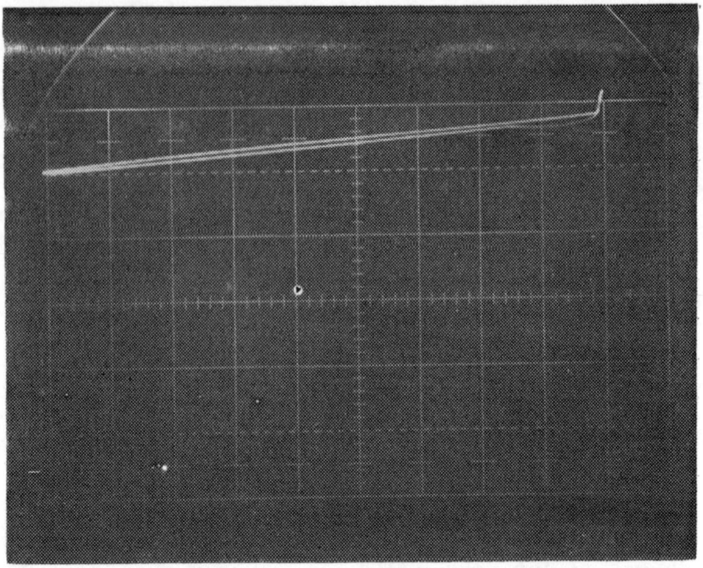

Fig. 5-12. A measurement of I_{CEO} on a 2N398A germanium PNP transistor. Curve-tracer settings are: collector series limiting resistance, 1000 ohms; horizontal sensitivity, 2V/cm; vertical sensitivity, 0.5 mA/cm. The step generator is not used, as the base lead is open. Germanium transistors will generally have higher leakage currents than silicon. Leakage on this particular device is 500 μA with a collector–emitter voltage of 18V.

It is important to make voltage breakdown measurements as rapidly as possible, or excessive power dissipation in the transistor may result from the high collector currents and voltages used in the procedure. Excessive power dissipation in a transistor may result in permanent device damage.

Collector-to-base and base-to-emitter breakdowns may be determined by connecting the base–collector or base–emitter terminals of the transistor to the curve tracer in the same manner as an ordinary 2-terminal diode would be checked for reverse breakdown characteristics. The unused lead of the transistor is not connected.

Leakage Current

If a transistor is to be used in relatively high impedance circuits, the amount of leakage current that the device exhibits may be of some importance. This leakage current is often specified in two different forms: I_{CEO}, which is the collector–emitter current with the base open; and I_{CES}, the collector–emitter current with the base shorted.

To measure I_{CEO}, the collector and emitter leads are connected to the curve-tracer terminals and the base lead is left open. The sweep voltage is then increased as far as possible, without exceeding breakdown, and the leakage current is measured as shown in Fig. 5-12. If the leakage current is quite small, it may be necessary to use the magnifier on the curve tracer to see this current. Frequently it will not be measurable at all. If the specification of I_{CES} is to be determined, the base lead is simply shorted to the emitter lead and the measurement is performed in the same manner as it would have been for I_{CEO}.

Saturation Voltage

Transistors that are used in digital circuits have a few additional special characteristics. When a transistor is used a a switch, the collector-to-emiter voltage when the transistor is *saturated* (that is, drawing its maximum collector current) is of great importance. Suppose, for instance, that a transistor is to be used in the common-emitter configuration to drive the input of a 7400-series TTL gate package. Specifications for 7400 logic require a logic 0 of at least 0.4V be maintained. This means that the transistor, when turned on, must have a collector–emitter voltage of less than 0.4V when sinking a maximum current of 1.6 mA from the input of the gate plus its own collector current.

To determine whether this transistor will properly drive its gates, a curve tracer may be employed to measure $V_{CE(SAT)}$, or the collector–emitter saturation voltage. Figure 5-13A shows the curves of an NPN device (2N2369), which is often used for such applications. Note that the saturation voltage increases as the collector current increases. Figure 5-13B shows the same transistor, but the base current has been increased by a factor of 100.

Temperature Effects

In certain critical applications the effects of temperature on transistor characteristics may be quite important. One way to ascertain that a transistor will or will not cause problems in a circuit is to place the transistor in operation in a curve tracer, using collector voltages and currents that will be encountered by the device in normal operation. Once a family of curves has been established, the device can be heated (a soldering iron, or even finger heat may be sufficient), and variations in these curves can be noted. With looping, shown in Fig. 5-14, the characteristics of the transistor have actually changed from the time the sweep made the deflection from the left to the right side of the screen to the time of its return.

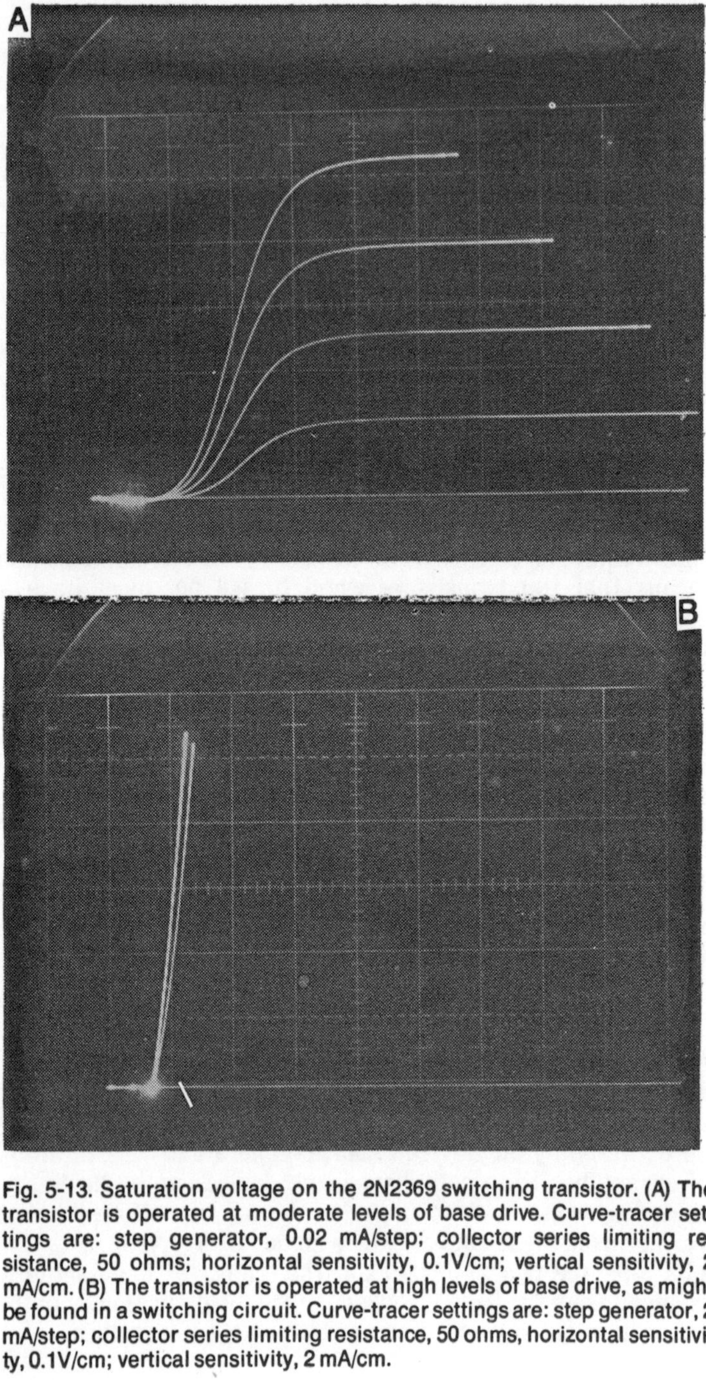

Fig. 5-13. Saturation voltage on the 2N2369 switching transistor. (A) The transistor is operated at moderate levels of base drive. Curve-tracer settings are: step generator, 0.02 mA/step; collector series limiting resistance, 50 ohms; horizontal sensitivity, 0.1V/cm; vertical sensitivity, 2 mA/cm. (B) The transistor is operated at high levels of base drive, as might be found in a switching circuit. Curve-tracer settings are: step generator, 2 mA/step; collector series limiting resistance, 50 ohms, horizontal sensitivity, 0.1V/cm; vertical sensitivity, 2 mA/cm.

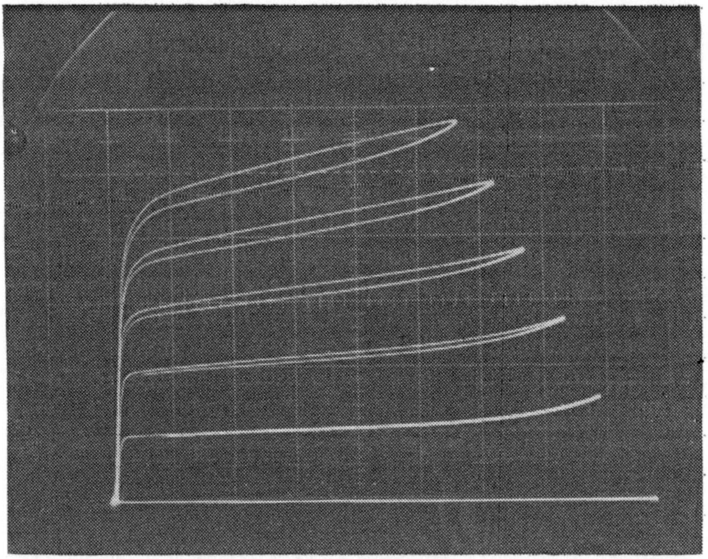

Fig. 5-14. Thermal looping caused by internal heating of the transistor. The transistor is a 2N3393, which has 200 mW dissipation. The curve-tracer settings are: step generator, 0.1 mA/step; collector series limiting resistance, 100 ohms; horizontal sensitivity, 5V/cm; vertical sensitivity, 20 mA/cm. Note that the point at the center of the display, where there is distinct thermal looping, represents a collector—emitter voltage of 20V at a current of 60 mA. This yields a power of 1.2W.

Thermal Runaway

Noting Fig. 5-15, one can see that with increasing collector voltages this effect can be carried to the limit; the curves shown in the figure represent thermal runaway. At this stage the transistor parameters are no longer under the control of the base current; they are completely out of control because of device heating. A transistor subjected to this test may be damaged quite rapidly.

General Testing Notes

There are a few general notes that should be made with respect to testing transistors on a curve tracer. First, setups should be made carefully, always starting with low initial voltages and currents, as semiconductor devices are easily damaged. Curve tracers are capable of producing large voltages and large currents, especially when compared to the limits on small-signal devices. Devices will be destroyed within the first few sweeps of an improperly set curve tracer.

Second, certain hazardous conditions are present during device measurements. Transistors can get very hot when their

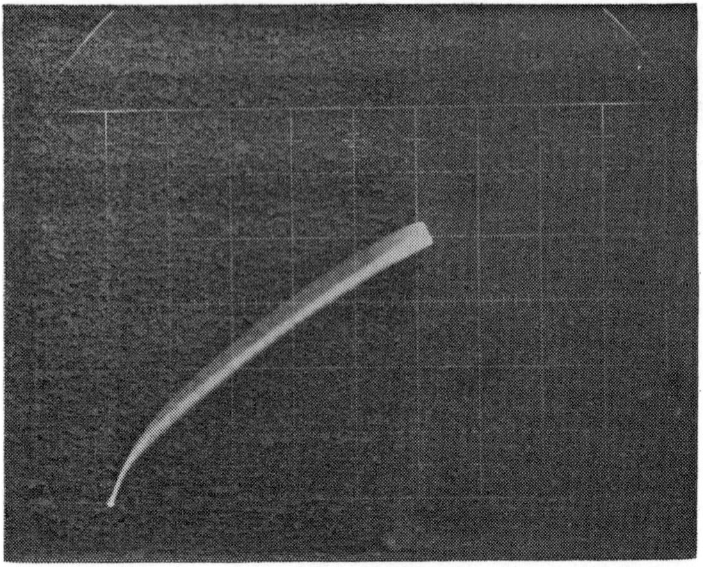

Fig. 5-15. Thermal runaway caused by excessive internal heating of the transistor. As can be seen, the characteristics of this transistor were changing as this photograph was being taken (1 sec exposure). The transistor was destroyed shortly after exposure was complete. Curve-tracer settings: step generator, 0.1 mA/step; collector series limiting resistance, 50 ohms; horizontal sensitivity, 5V/cm; vertical sensitivity, 50 mA-/cm.

parameters are exceeded, and severe burns can result from contact with the cases of both power transistors and small-signal transistors which have been used beyond their limits.

Third, curve tracers also have the ability to produce exceptionally high voltages to insure breakdown of most junctions; they also have fairly large current capabilities. The net result is power supplies which are lethal. It is altogether too easy for a person to come in contact with leads on devices that are under test, especially those which have metal cans. *Be sure to reduce the sweep voltage to zero prior to connecting or disconnecting the device from its socket.*

Fourth, when intermittent tests for device breakdown are desired, the A-B switch is an excellent method of rapidly controlling the curve tracer. In comparing devices, a standard device can always be left in one socket, and the second socket can be used to interchange devices that are being compared or sorted.

Fifth, the user must remember when making in-circuit tests that parameters may not be exactly as expected because of other components in the circuit, such as resistors and large capacitors. To make exact parameter measurements on a transistor, it is best to remove it from the circuit; however, general indications as to its characteristics, relative gain, etc., may be made while the device is in the circuit. Power should always be removed from a circuit prior to making in-circuit tests.

FIELD-EFFECT TRANSISTORS

Like the bipolar transistor, the field-effect transistor has two major categories of device polarity. Unlike the bipolar device, however, the field-effect transistor is a voltage-operated device, so initial setups for measuring field-effect devices include changing the step generator from the current mode. At present there are six major categories of field-effect devices, and their setups are shown in Table 5-1. A special note should be made of enhancement-mode field-effect devices, as certain curve tracers are not able to achieve the required combination of sweep polarity and step-generator-voltage polarity to properly display their characteristics. One method to overcome this—for example, with the N-channel enhancement-mode device—would be to use a positive collector sweep voltage and a negative step voltage. This step voltage must be offset by some positive voltage so that the most negative step produces a zero gate-to-source voltage. One technique for achieving this is to place a battery or an adjustable power supply in series with the gate (the base terminal) of the curve tracer.

Transconductance

As with the bipolar transistor, one of the most important measurements that may be made is that of device gain. The gain of the field-effect transistor is referred to as its *transconductance* (g_m). Transconductance is a ratio of the change in drain current caused by a change in gate voltage and is expressed in micromhos. Figure 5-16 is a display of an N-channel junction field-effect device (2N4416). To make a transconductance measurement, a change in drain (sweep) current is read from the curve tracer. Because the step value causing this current change is known, the change in drain current may be divided by the change in gate (step) voltage, yielding transconductance.

Note that the transconductance is not uniform from curve to curve, and therefore the transconductance must be measured at or about the current that the FET will normally be used. Once the gain characteristics have been determined,

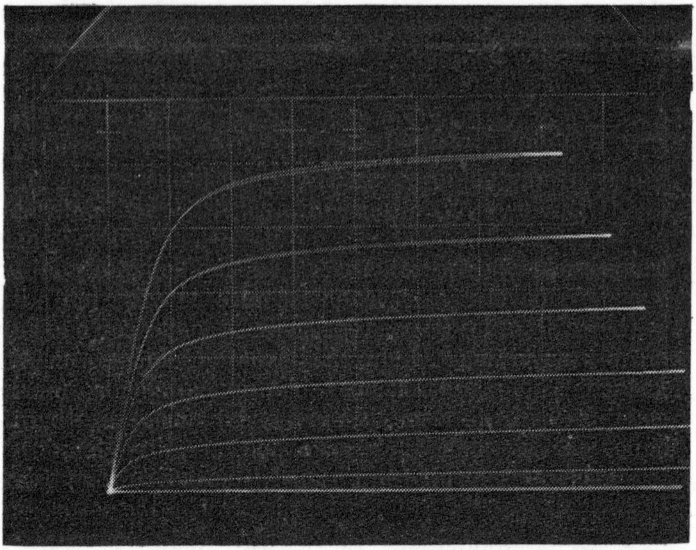

Fig. 5-16. A family of curves for the 2N4416 N-channel JFET. The uppermost curve is the curve for step generator output of 0V; succeeding downward curves represent increased voltage output from the step generator. Curve-tracer setting are: step generator, −0.5V/step; drain limiting resistance, 500 ohms; horizontal sensitivity, 2V/cm; vertical sensitivity, 2 mA/cm.

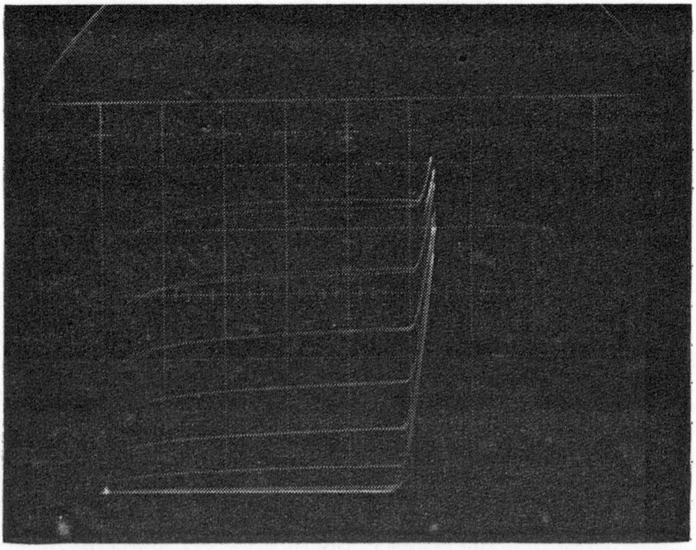

Fig. 5-17. Breakdown characteristics of a 2N4416 N-channel JFET. Curve-tracer settings are: step generator, −0.5V/step; drain limiting resistance, 500 ohms; horizontal sensitivity, 10V/cm; vertical sensitivity, 2 mA/cm. The drain-to-source breakdown of the 2N4416 is specified as 30V.

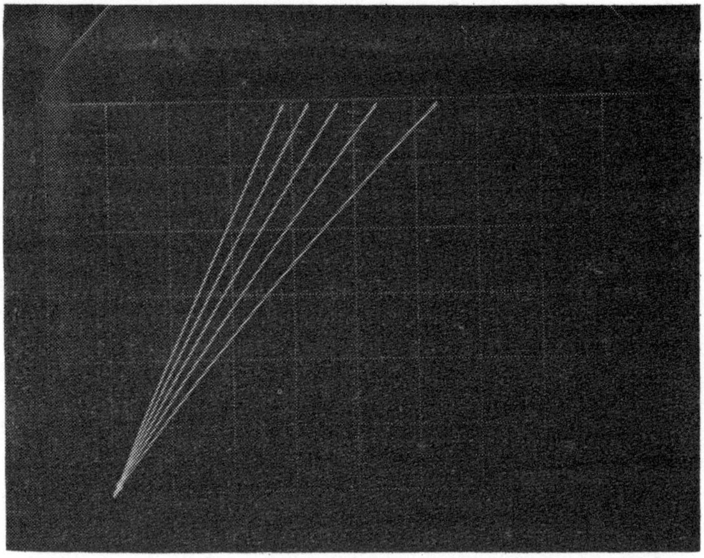

Fig. 5-18. The on resistance of a TI TIS-73 N-channel JFET. This device is especially designed for switching applications. The curve-tracer settings are: step generator, −1V/step; drain series limiting resistance, 10K; horizontal sensitivity, 0.02V/cm (0.1V/cm with a ×5 magnifier); vertical sensitivity, 0.5 mA/cm. The steepest curve (the first) has a slope of 24 ohms. This curve is generated by a drain sweep with the gate-to-source voltage at zero. Specifications for the TIS-73 call for 25 ohms maximum resistance.

the next most important characteristic is voltage breakdown. Figure 5-17 shows the breakdown characteristics of the 2N4416 and, as you can see, the characteristic curves are quite similar to those of the bipolar transitor.

"On" Resistance

Field-effect transistors that are used in digital applications or for switching analog signals have another important characteristic, the device *on* resistance, expressed as R_{on}.

With the N-channel JFET, this is the resistance from drain to source with the gate-to-source voltage at zero. The value of R_{on} may be determined as shown in Fig. 5-18 by increasing the horizontal sensitivity of the curve tracer until the resistance of the *on* portion of the curve may be measured. The *on* resistance of a device is determined in a manner that is similar to a measurement of the dynamic resistance of a diode.

Temperature Effects In FETs

Temperature effects on the characteristics of an FET may be made in the same manner as for a bipolar device. Special precautions that should be taken when testing field-effect transistors include taking extreme care with MOS devices, especally those that are unprotected, to make sure that the terminals are not exposed to static electricity. One method to achieve protection is to wrap a fine wire around the leads of the transistor before removing it from the circuit, and then unwrapping this wire when the device is securely mounted in the curve tracer. If the device is a dual-gate MOSFET the unused gate should either be returned to the source terminal of the device or, if the characteristics are specified with a certain gate-to-source voltage on the unused gate, this should be supplied from an external source. In no case should MOS devices be tested with the gates left unconnected.

SPECIALTY DEVICES

The characteristics of the silicon controlled rectifier (SCR) and the dual version of the same, the triac, may be easily examined on the curve tracer.

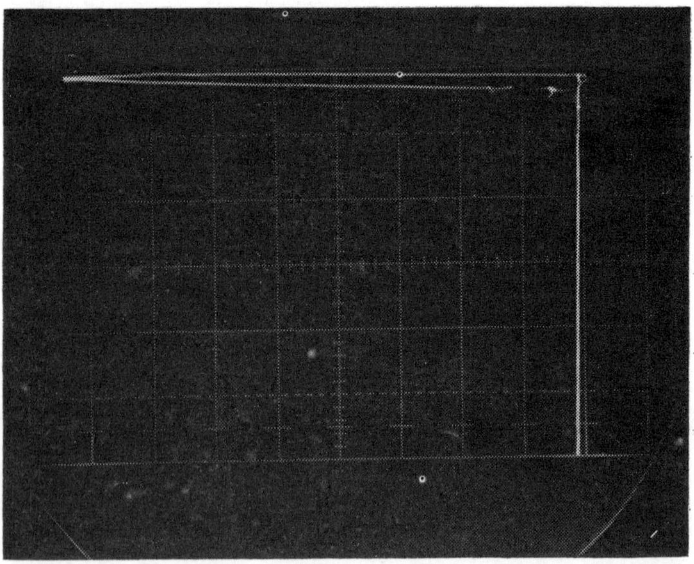

Fig. 5-19. The forward characteristics of a TIC-44 SCR. The curve-tracer settings are: anode series limiting resistance, 50K; horizontal sensitivity, 40V/cm (20V/cm with a ×0.5 magnifier); vertical sensitivity, 0.5 mA/cm. Note the forward blocking voltage of this device is 230V, and the holding current is 0.5 mA.

SCR

Figure 5-19 shows the curves of the SCR where the device characteristics of forward blocking voltage and holding current are to be measured. As with the diac, forward blocking voltage is the maximum collector or horizontal voltage that can be reached, and the holding current is the lowest forward current that can be maintained. When these measurements are made, the step generator is maintained at zero current output.

Figure 5-20 shows a display to measure gate trigger current and *on* resistance of the SCR. To measure the gate trigger current, the collector sweep voltage is reduced to a point below the forward blocking voltage of the SCR. The STEPS PER FAMILY control is adjusted for a known number of steps—if necessary, this is the maximum number of steps available from the curve tracer— and the STEP RANGE control is advanced from the smallest step current to the first postion at which the vertical line appears. The gate sensitivity is then known to lie between the position just prior to this and the postion which caused the vertical line.

Once a vertical line has been achieved, the *on* resistance of the SCR may be measured by expanding the horizontal sensitivity and computing the resistance represented by the vertical line.

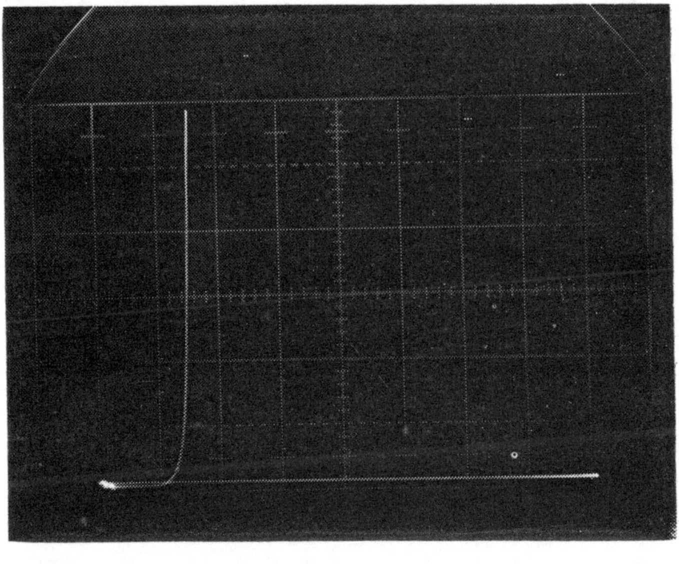

Fig. 5-20. Measurement of the on resistance of the TIC-44 SCR. Curve-tracer settings are: anode series limiting resistance, 1000 ohms; horizontal sensitivity, 1V/cm; vertical sensitivity, 1 mA/cm.

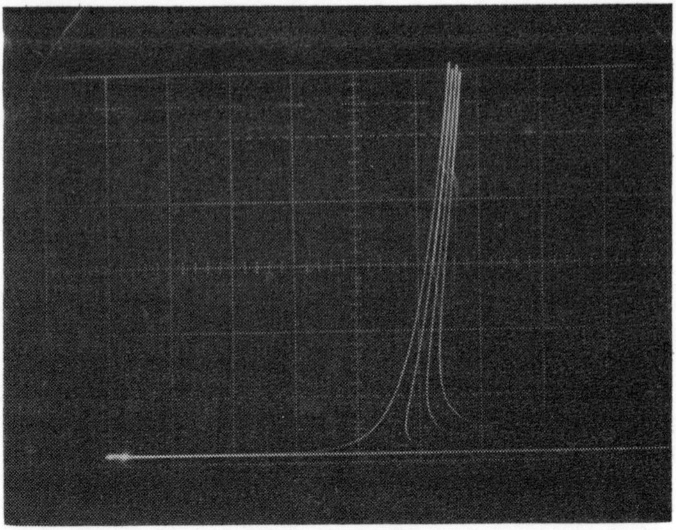

Fig. 5-21. Measurement of the interbase resistance of the 2N2646 unijunction transistor. The unijunction is connected to the curve-tracer terminals as follows; B_1 to E, B_2 to B, and E to C. The curve-tracer settings are: step generator, 0 mA/step; emitter series limiting resistance, 500 ohms; horizontal sensitivity 0.2V/cm; vertical sensitivity, 0.5 mA/cm.

With the SCR, the reverse blocking voltage is essentially the reverse voltage of the main-diode portion of the SCR. This measurement may be made in exactly the same manner as the reverse breakdown voltage of a diode.

Forward measurements on a triac will be made the same way as those of the SCR. Reverse measurements will give curves identical to the forward set but inverted, as the triac acts like two SCRs in parallel, with one reversed.

Unijunction Transistors

Figure 5-21 indicates the characteristics of a unijunction transistor (UJT) connected as indicated in Table 5-1. The slope of this curve gives a measurement of the inner-base resistance of the device.

SUMMARY

The number of devices that can be analyzed by the curve tracer is limitless. All measurements that have been previously discussed can be applied on one form or another as in-circuit measurements as well as measurements on devices. When making in-circuit measurements, make sure that the power supplies are turned off and, if possible, disconnected

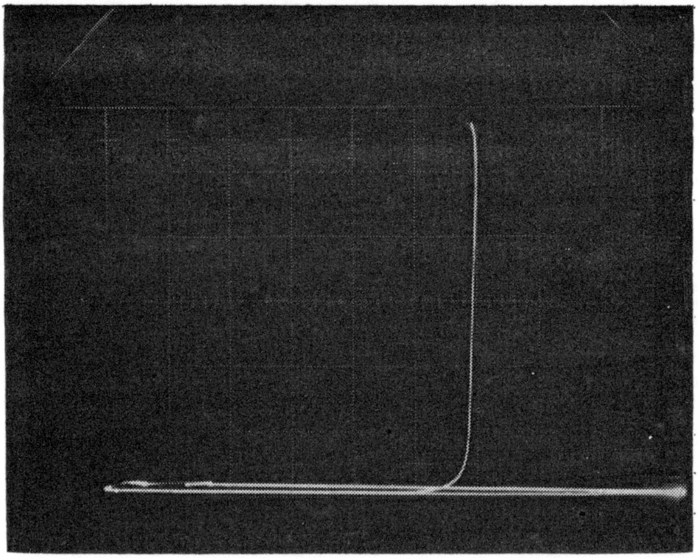

Fig. 5-22. Measurement of the breakdown and holding voltage of a NE-2H neon lamp. Curve-tracer settings are: series limiting resistance, 10K; horizontal sensitivity, 20V/cm; vertical sensitivity, 1 mA/cm. For this particular lamp, breakdown occurs at 90V, and holding voltage is 60V at 1 mA.

from the device under test. And remember that resistances in the circuit in addition to those of the device may cause some difference between the characteristic curves obtained and those expected.

Virtually any device an be measured with the curve tracer—neon lamps (Fig. 5-22),tungsten lamps, LEDs, vacuum tubes—even integrated circuits can yield characteristic traces, which may tell a great deal about their function. Because of its fundamental measuring concepts, it is likely that the curve tracer will be directly applicable even to new devices as they are developed.

Chapter 6
Test Leads and Probes

It is difficult to look on a 3 ft length of RG-58/U cable as an instrument, but it would be even more difficult to use any instrument if there were no probes, cables, connectors, or terminators. The theory, specifications, and operating techniques are the same for these devices when used with any instrument discussed in this book.

FUNDAMENTALS OF INSTRUMENT CONNECTIONS

Devices which interconnect instruments and devices under test may be divided into three categories: test leads, transmission lines, and probes. Each of these will be examined in this section so that the fundamental concepts are clear and the basic dividing lines between one type of interconnection device and another are established.

Test Leads

Test leads are the simplest form of interconnection possible. In many cases they consist of nothing more than a single-wire conductor for the signal and another wire for the return. Such leads are common on DC voltmeters and on AC voltmeters used to measure power-line-frequency signals. Frequently only one wire is apparent, as the common ground between the instruments acts as the return path. This system is quite acceptable so long as the signals are of high amplitude and low frequency and are from low-impedance sources.

It is important to remember that the return path is always necessary. Often, when using high-frequency circuits, a single lead will appear to be the only point of connection. Don't be misled: Electricity is still electricity, and return paths are required. The path taken for the return signal may well modify the measured signal considerably. It is always wise to determine just what the return path is.

When the source impedance is no longer low or the signals are no longer high level, then interfering signals may develop on a simple open-wire test lead. The solution to this problem may be a twisted cable. Interference with signals on the simple open-wire test lead may be caused by the signal and common wire being subjected to uneven electrical fields (see Fig. 6-1). One of the attributes of the twisted cable is that, with the two leads in close proximity, the electric fields will have an equal effect on both conductors and therefore cancel out. As shown in Fig. 6-2, the capacitive divider between the interfering source and the signal wire will develop an interfering signal.

If the signal wire is completely surrounded by a second conductor at ground potential, a different situation develops (see Fig. 6-3). There can be no capacitive divider from the interfering source to the center conductor to ground if the shielding effect of the outer conductor is perfect. If there is no capacitive divider, there will be no interfering signal coupled to the center conductor. Most shields are not perfect, so there will be some signal coupled to the center conductor of the shielded cable. However, the amount coupled to the center conductor will be considerably less than without a shield.

Transmission Lines

The transmission line and shielded cable are similar in nature and often confused. Both frequently use a cable which has a single center conductor surrounded by a woven braid; a coaxial cable. When a coaxial cable is used as a transmission

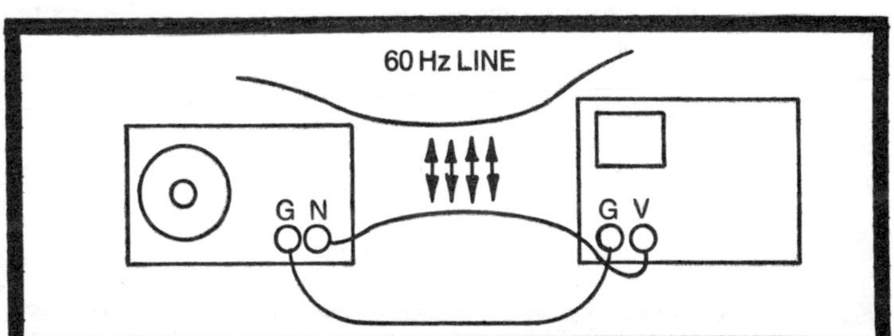

Fig. 6-1. An interfering signal is generated on one of a pair of test leads used to connect an oscilloscope to a generator. This effect can be reduced to some extent by bringing the two test leads into close proximity.

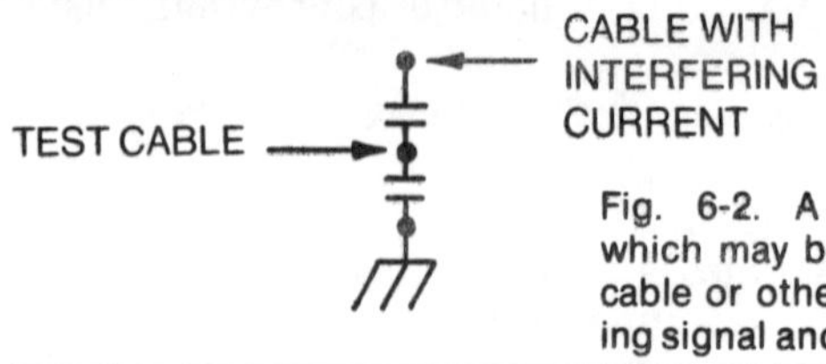

Fig. 6-2. A capacitive divider which may be formed between a cable or other source of interfering signal and a test cable.

line rather than as a shielded cable, the ability to deliver power and the characteristic impedance of the cable become of great importance.

If AC electrical energy is applied to a transmission line, that energy will assume a particular voltage/current ratio as the energy is propagated over the length of the line. This effect is caused by the distributed capacitance and inductance of the cable, which are determined by the mechanical dimensions of the line. As a voltage/current ratio has units of resistance, this natural voltage/current ratio of the transmission line is referred to as the *characteristic impedance* of the transmission line.

If a coaxial transmission line is terminated in a resistance (load) that will maintain the voltage/current ratio caused by the characteristic impedance of the line, all the transmitted power will be absorbed by the load. If, however, the resistance of the load differs from the characteristic impedance of the transmission line, some of the power will be reflected back along the line. Reflected power results in *standing waves*, which occur when the reflected signal adds and subtracts with the forward signal. Voltage and current maxima and minima occur at wavelength multiples on the line.

One of the most frequent effects of standing waves is an apparent signal loss to or from the instrument in use. If a

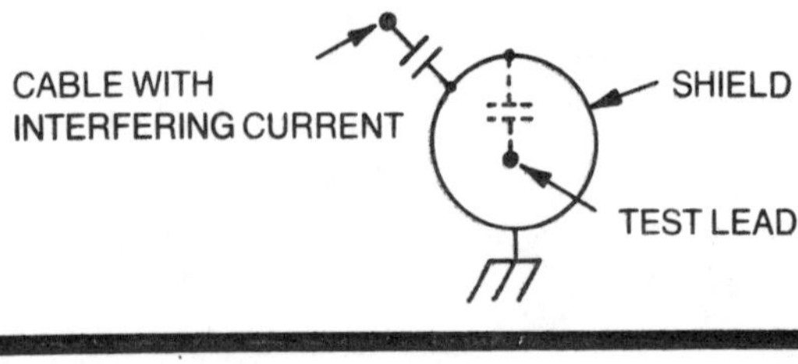

Fig. 6-3. The effect of placing a shield around a test cable. The capacitive divider that once existed from the interfering source to the test cable to ground no longer exists, as all paths to the test cable from the interfering source are shielded.

voltage minimum occurs at a point where an instrument is connected to the cable, a voltage of much less amplitude than intended is presented to the instrument, which then appears to lack sensitivity.

The opposite can also happen, and too great a signal may be applied to the instrument. Neither of these conditions will occur if the line is *flat*, that is, terminated in a resistance equal to its characteristic impedance.

Another problem occurs when the coaxial transmission line is being driven with a square wave or some other form of pulse. If the line is not terminated in its characteristic impedance, the reflections may produce additional pulses on the line. If an instrument such as a frequency counter is connected to this line, it may count these extra pulses, resulting in an erroneously high displayed frequency.

If the transmission line is not driven by a source having an output impedance equal to the characteristic impedance of the transmission line, two problems may occur. First, there will not be maximum power transfer to the line nor therefore to the load. (For example, a 600-ohm generator cannot supply maximum power to a 50-ohm system.) Second, reflections will occur. These reflections will be propagated back to the generating source. If the generating source does not present an impedance equal to the characteristic impedance of the line, the reflected power will not be absorbed at the generating end, and further reflections will occur. This effect will increase the standing waves and the probability of measurement error.

The problem of power transmission is important if a system of coaxial transmission lines is used to conduct signals about an instrumented experiment. These lines must be properly terminated to prevent error-causing reflections, and a low-impedance system will be necessary. This impedance will usually lie between 50 and 93 ohms. Say, for example, a 1V signal is to be transmitted from one location to another and must be maintained across 50 ohms. By the formula $P = E^2/R$, we can see that 20 mV of power is involved. Compared to the power involved in a 1M system (1 μW), this is considerable power. To supply a 1V signal, the generator must have at least 20 mW of power-generating capability. If we desire 10V on a 50-ohm system, the generator must be capable of $10^2/50$ or 2W of power.

LOW-CAPACITANCE PROBES

At first, probes were nothing more than a simple method of attaching a test lead to a circuit. The test probe supplied with a VOM retains this original simplicity. But as measurement techniques have improved, the simple probe has proved to be no longer effective in certain measurement applications.

One example is that of the simple probe attached to the end of a shielded cable. This combination provides a convenient method of transmitting interference-free signals from the point of measurement to the instrument, and at low frequencies such a system is quite acceptable; but as the measured frequency is increased, the capacitance of the cable begins to have a serious loading effect on the circuits being measured.

For example, consider a simple measurement in which a source with an effective impedance of 1K and a frequency of 1 MHz is to be measured. A 3 ft length of RG-58/U coaxial cable is used in conjunction with an oscilloscope having an input impedance of 1M shunted by 30 pF. The total capacitance at the source will be 30 pF from the oscilloscope plus an additional 85.5 pF from the RG-58/U (28.5 pF per foot). If the effects of the capacitance are not taken into account, the loading caused by the 1M oscilloscope on the 1K source impedance of the generator will cause less than 0.1% error (Fig. 6-4).

However, the capacitance cannot be ignored. The total shunt capacitance is 115.5 pF. At 1 MHz this has an impedance of 1378 ohms reactive. The 1M of the scope is so dwarfed by the low capacitive reactance that it may be left out of the picture entirely. A simple calculation (see Fig. 6-5) shows us that a

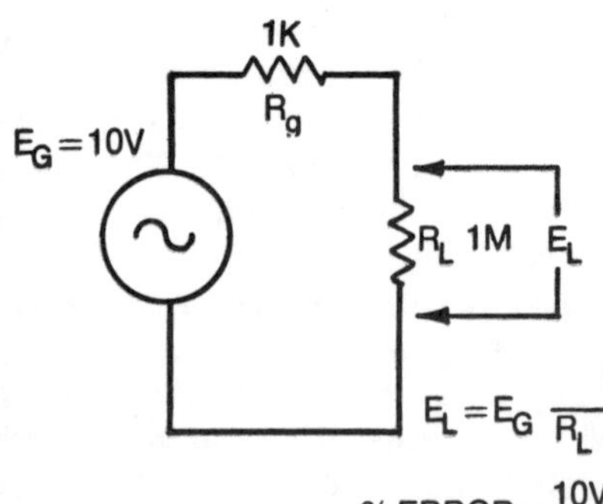

$$E_L = E_G \frac{R_L}{R_L + R_G} = \frac{1,000,000}{1,001,000} = 9.99V$$

$$\%\,ERROR = \frac{10V - 9.99V}{10V} \times 100\% = 0.1\%$$

Fig. 6-4. The error caused by a 1M load placed on a generator with 1K source impedance. Resistance R_L is the 1M input impedance of the instrument.

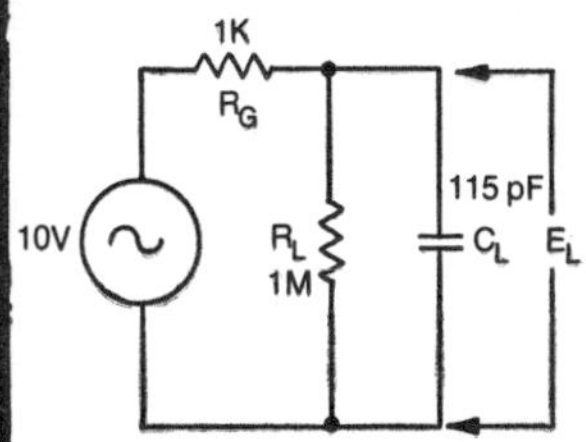

$$X_C = \frac{1}{2\pi fC} = \frac{1}{2 \times 3.14 \times 10^6 \times 115.5 \times 10^{-12}} = 1378\Omega$$

$$E_L = \frac{X_C}{\sqrt{(R_G)^2 + (X_L)^2}} = \frac{1378}{1703} = 8.1\text{V}\angle 39^\circ$$

$$\%\text{ERROR} = \frac{10\text{V} - 8.1\text{V}}{10\text{V}} \times 100\% = 19\%$$

Fig. 6-5. The error caused by a 1M load shunted by 115.5 pF placed on a 10V, 1 MHz generator with 1K source impedance. Error calculations disregard the 1M, as its contribution to the total error is insignificant.

resistive–capacitive voltage divider has been formed, and the 10V signal at the generator is now presented to the oscilloscope as a signal of 8.1V, phase shifted by 39°. All this at only 1 MHz.

If the signal frequency is increased to 10 MHz, the cable capacitance now presents a low 138 ohms reactive. The voltage presented to the oscilloscope is reduced from the actual 10V of the source to less than 1.4V by loading of the source impedance with the cable capacitance. In other words, the amplitude error at 1 MHz is 19%, and at 10 MHz it is greater than 86%.

Obviously such error cannot be tolerated. It is also obvious that all capacitance in the system cannot be eliminated, nor can all the source impedance be eliminated. One method of combating such error is to attempt to measure the signal at known low-impedance points. This is not always possible, and as the measurement frequency increases, even low-impedance sources will be affected by capacitive loading.

The passive probe is used to reduce the measurement capacitance. Placing two capacitors in series results in a total capacitance lower than the smaller capacitance of the two. This principle makes capacitance reduction through the use of the passive probe possible. Figure 6-6 diagrams the effect of the passive probe. A second capacitor that is one-ninth the capacitance of the cable and the oscilloscope combined is placed in series with the signal being supplied to the cable. The results of this added capacitor are threefold: First, the capacitive loading on the source is now less than one-tenth the value of the cable and scope alone. Second, the two capacitors act as a 10:1 voltage divider with a division ratio which is constant at all frequencies so long as the ratio of the two capacitors remains the same and the effect of the 1M shunt resistance is negligible. Third, the probe is no longer useful at DC.

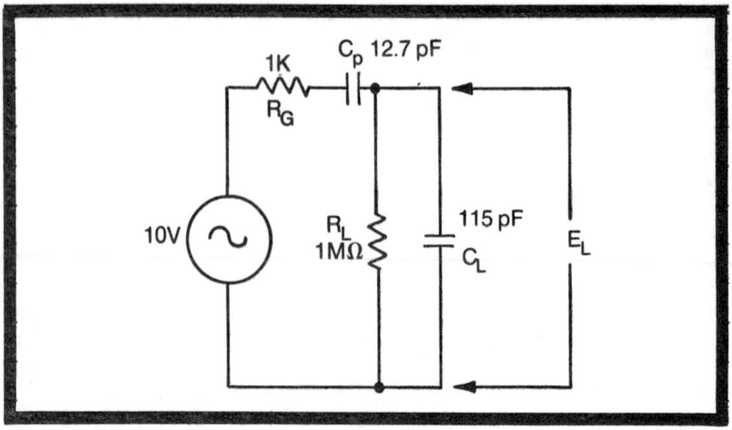

Fig. 6-6. The reduction in measurement capacitance by the insertion of a probe series capacitor at the probe tip. The probe capacitance results in a smaller load capacitance being placed on the generator and creates a capacitive voltage divider.

For the final reason the oscilloscope probe is modified as shown in Fig. 6-7. A resistor is added in parallel with the series capacitor. This resistor and the 1M resistance of the oscilloscope form a low-frequency-and-DC voltage divider having exactly the same ratio as the capacitive divider. A variable shunt capacitor is usually added across the probe output so the capacitive division ratio may be adjusted to exactly match the fixed resistive ratio. This permits one probe to match instruments with different input capacitances.

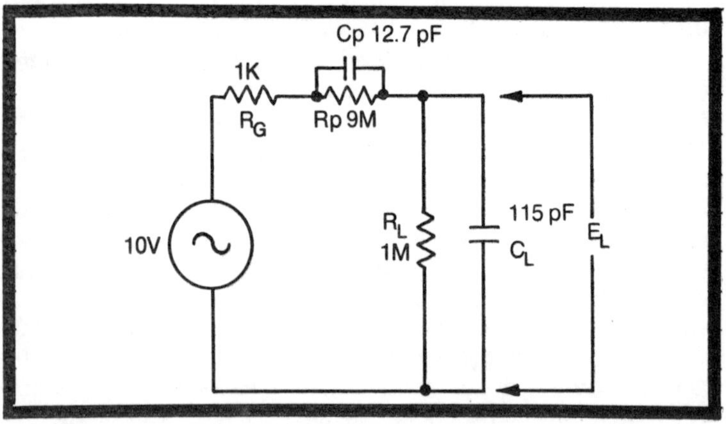

Fig. 6-7. The passive low-capacitance probe. The voltage division of the probe is constant for all frequencies, including DC. Often a portion of C_L is made variable so the ratio of the capacitive divider may be matched with the ratio of the resistive divider.

The center conductor of the cable is usually made of a resistive material. The resistive cener conductor dampens ringing associated with a coaxial cable which is not properly terminated in its characteristic impedance.

Passive probes usually come as decade dividers. The ×10 probe is by far the most popular. Occasionally a ×100 probe is desired, as input capacitance is further reduced and a 100M input impedance is achieved. Of course, this is accompanied by an attenuation factor of 100, which tends to desensitize all but the most sensitive measurement equipment. Occasionally a division factor of 50 will be found in a passive probe. Note that while the probe acts as a voltage *divider*, the name is given as a *multiplier*. A ×10 probe *divides* the input signal voltage by a factor of 10. The ×10 term indicates the multiplication factor that must be applied to a displayed signal to know the correct amplitude.

OTHER PROBES

Other probes are available for special use. The RF and demodulator probes have a great deal in common. Both of these are specialized probes using diode circuits to rectify RF signals.

A simple RF probe is shown schematically in Fig. 6-8. Radio-frequency signals present at the input of the probe are rectified by the diode. The resultant DC voltage is filtered by the capacitor. The probe may also contain some series resistance which will, when the probe is used with a meter of known input resistance, create a voltage divider. This voltage divider will divide the voltage stored on the capacitor (which will be equal to the peak value of the RF signal) by 1.414 to yield a meter deflection that may be read as RMS voltage.

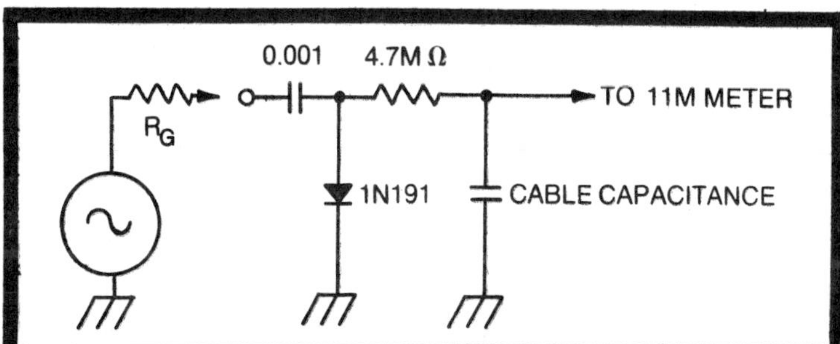

Fig. 6-8. A schematic of a typical RF probe designed for use with a VTVM having 11M input impedance. The voltage divider created by the 11M meter input impedance and the 4.7M series probe impedance makes the peak-to-RMS correction.

Additional diodes may be connected in series to increase the reverse breakdown voltage of the diode and therefore the maximum operating voltage of the probe. Placing additional diodes in series increases the minimum voltage that can be measured by the probe.

The *demodulator* probe is similar in design to the RF probe but usually differs in its intended use and output time constant. Demodulator probes, as their name would imply, are designed to recover AM information on an RF signal. These probes are frequently used in analyzing television signals in such places as the IF strip and in recovering signals created by a sweep generator. Signals from a sweep generator often obtain modulation from the amplitude–frequency characteristics of the circuit being swept. Demodulator probes are primarily used with oscilloscopes, RF probes are used with voltmeters.

Other types of specialty probes include those which are used to measure current; *active* probes containing amplifiers to achieve high-impedance at no sacrifice in gain; *logic* probes (which are really miniature instruments rather than probes) for detecting "high" and "lows" in logic circuits; and specially built *high-voltage* probes containing series multiplier resistors to permit extremely high DC voltage measurements with ordinary meters such as VTVMs and VOMs.

CONNECTORS

A review of the hardware used with probes and cables is important because, although much of this hardware is common and therefore familiar to most users, there is a great amount of hardware that has been expensive for many years and is just now coming into common use.

Shown in Fig. 6-9 are a number of connectors used with DC and low-frequency AC test instruments today.

The simple banana plug is still the most commonly used connector for high-level DC and low-frequency AC instruments. Such items as voltmeters, low-frequency generators, and a few oscilloscopes use this connector. The banana plugs come in two sizes—regular or ¾ in., and miniature, or ½ in., designated by the spacings given to dual connectors. Banana plugs are usually given a rating of 10A to 15A DC current-carrying capacity. Both sizes come in the stackable and nonstackable configuration. For test leads, the stackable ones are especially handy.

The dual banana plug has two connectors spaced on ¾ in. centers, still the most common distance. Dual banana plugs are most commonly used in connecting shielded cables to

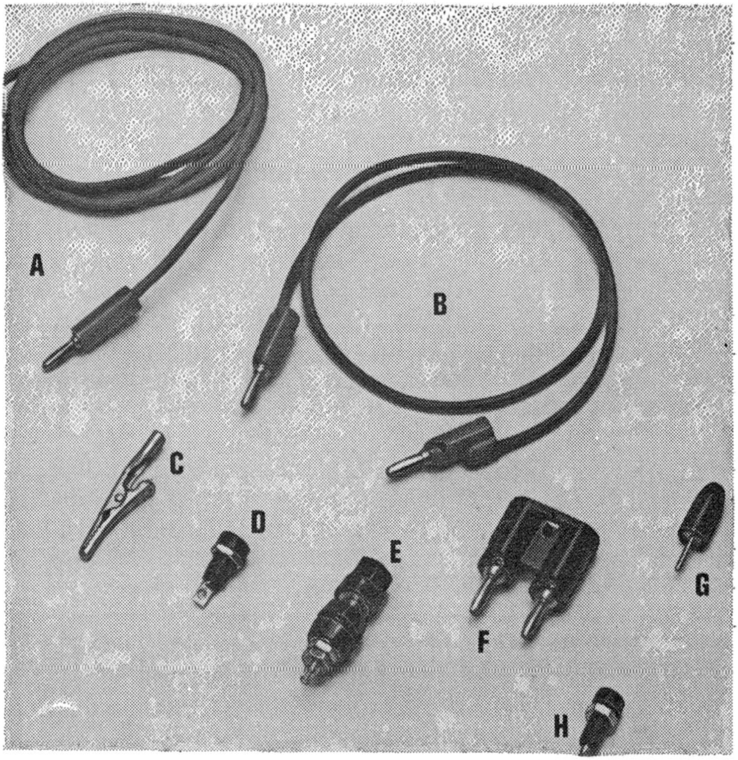

Fig. 6-9. Some of the low-frequency test cables and connectors in common use today. (A) The test probe terminated in a banana plug, commonly used as input cable for a voltmeter. (B) The banana-to-banana test or patch cable. This cable has stackable banana plugs, especially handy when interconnecting grounds and in power distribution. (C) The alligator clip. This clip can fit over a banana plug. There are clips available with threaded ends for attachment to test probes such as one in A. (D) The panel-mount banana jack. (E) The 5-way binding post. This jack is very common on the outputs of low-frequency generators and on power supplies. (F) The dual banana plug, often used to terminate shielded cables on low-frequency generators and AC VTVMs. Note the rib on the left side, which is used to identify the ground side of the connection. (G) The tip plug, common on earlier VOMs due to its compactness. (H) The panel-mount tip jack.

banana jacks. Most instruments have a set of banana jacks mounted on ¾ in. spacings to accept such a dual plug. Often the dual plug will have a rib on one side so the shield or ground lead of the test cable may be easily identified.

Connections for the banana plug come in two forms—the simple banana jack and the 5-way binding post. The simple banana jack has a capacity of 2–5 pF to the metal panel. The

5-way binding post has a similar capacity to the panel upon which it is mounted, and it will also accept wire, spade lugs, tip jacks, or alligator clips. The 5-way binding post offers the additional advantage of making a very low resistance connecton when the screw cap is tightened onto a wire, tip plug, or spade lug. Five-way binding posts come in both regular and miniature sizes (¾ and ½ in.).

Some voltmeters use a much smaller connector than the banana jack, called the *tip* jack. In the tip connector the jack is a solid metal part, while the banana jack consists of four spring leaves, which are bowed to make a firm connection in a round hole. Connection is made by spring elements in the tip jack. Tip jacks are rated at currents and capacitances similar to those of the banana jack.

Banana jacks and plugs, 5-way binding posts, and tip jacks and plugs should not be used for measurements exceeding the 1500V–2000V range for DC and line-frequency measurements. Special high-voltage connectors are made for such work and prevent the hazardous conditions presented when tip or banana jacks are used at these potentials. Measurements on signals such as the 15.75 kHz signals from TV high-voltage systems should not be made using these connectors.

Color coding for the binding-post caps as well as for test leads and other plug and jack covers is not standardized to any great extent; however, black is usually used for the common (negative, ground, or low) input terminal, and red is used for the "high" (positive) input terminal. Green is frequently used for the chassis ground terminal.

A triangular mounting configuration is often used for binding posts or jacks when the instrument is *floating*, that is, when the "low" or common terminal is not connected to chassis ground (see Fig. 6-10). A standard shorting link may be use to connect between ¾ in. or ½ in. spaced binding posts, such as the "low" common to ground. A short length of wire is also satisfactory.

Connections made where small potentials are measured prevents additional voltages which might be generated by the contact of two dissimilar metals. For this reason many of the 5-way binding posts are of high-quality metals when they are to be used on instruments capable of measuring or supplying small potentials. Some low-cost jacks and binding posts are made of aluminium. Mating aluminium plugs will generally not be found and are not desirable. Aluminium jacks should be avoided when low millivolt measurements are being made.

Shielded cables and terminated transmission lines should be interconnected with other equipment by coaxial connectors,

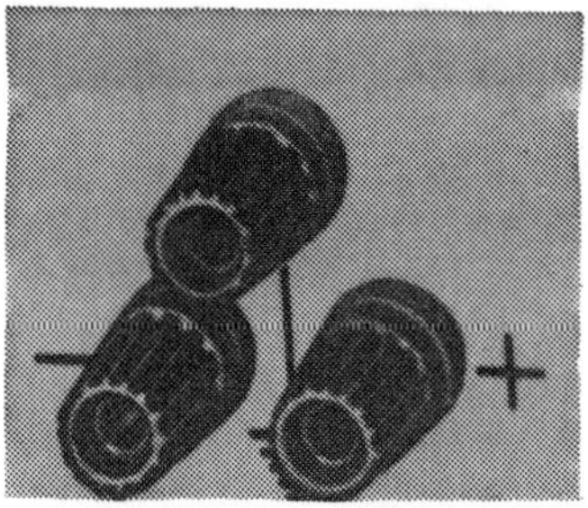

Fig. 6-10. The output terminals of the Heath IP-27 50V, 1.5A power supply. Note that the power supply may have either the positive or negative terminals connected to the earth ground, or the supply may be operated with neither terminal connected to earth ground. All three terminals are on a ¾ in. center-to-center spacing with all other terminals.

although the shielded cable is often terminated in the dual banana plug. Two types of coaxial connectors predominate. The most common is the *BNC connector*. This small connector, which readily connects and disconnects with a simple twist, is a natural for RG-58-sized cables. Until fairly recently, this connector has not been popular with the experimenter or serviceman as it has been quite expensive and very difficult to assemble at the end of a cable. Recent smplifications in the design of the BNC connector have reduced the cost and made field assembly and repair quite practical. The BNC connector is close to 50 ohms in characteristic impedance, and the common connector is good to 300 MHz or higher. Special BNC connectors may be obtained which can be used into the gigahertz region. BNC connectors are rated to 500V DC and 1500V AC (RMS). Most new oscilloscopes, RF generators, function generators, frequency counters, and pulse geneators use BNC connectors. Most adapters, attenuators, and terminators are available at their lowest cost with BNC connectors.

The old *UHF connectors* (SO-239 jack and PL-259 plug) are still found on some instruments. The disadvantages of the UHF connector lie in its size, the many turns of the screw cap for disconnecting or connecting, and the reflections when installed in a 50-ohm line. The UHF connectors have similar voltage ratings to those of the BNC type and are not recommended for use above 300 MHz. The UHF connector will fit larger coaxial cables, up to RG-8, which BNCs will not. This connector has been quite popular in HF and VHF communications equipment. It is wise to have a few BNC-to-VHF adapters available, especially if communications equipment is frequently worked on.

Occasionally when dealing with UHF and higher frequencies, it is necessary to connect to the larger cables, such as RG-8, or a better connector than either the BNC or UHF is required. The *type N* is often used in such instances. The type N connector comes in both 50- and 75-ohm models and is noted for its extremely low reflections. Type N connectors are difficult to assemble properly and command a high price. Frequently BNC- to-N adapters are used; however, such adapters are not completely reflection-free and should be kept out of exacting applications.

The *type F* connector is used with 75-ohm television RF transmission line. This connector makes use of the solid center conductor of the coaxial cable for the center conductor of the male connector. The main advantage of the type F connector is its low cost. Only instruments specifically designed for television RF use are found with the type F connector.

Three other coaxial connectors are occasionally found, usually on shielded cables rather than transmission lines. The *single button microphone* connector was chosen for a few instruments (it can usually be neatly replaced by a BNC jack, in the same hole) and at this time has little to recommend it. Although never really popular, because of their inexpensive construction, the *phono* plug and jack make a reasonably good connection for small coaxial cables. This connector has been used with reasonable success to 500 MHz and has voltage ratings that will permit use to 100V. The 2- or 3-conductor *phone plug* is sometimes found in audio equipment, and it is also used as a test lead connector on some VTVMs. Figures 6-11 and 6-12 show a few of these connectors.

COAXIAL CABLES

Three different coaxial cables are in common use for interconnections in instrumentation systems. These are RG-58, a 50-ohm cable, the most common; RG-59, a 75-ohm cable, most frequently found in television systems; and RG-62, a 93-ohm cable of much lower capacity per unit length than the other two. The essential characteristics of these three cables are listed in Appendix B.

Although most instrumentation work is done with the three previously mentioned coaxial cables, there are applications for other cables. Specialized applications include: low-cost microphone cable, in which neither capacitance nor characteristic impedance is important; a conductor pair inside a shield, where differential measurements are to be made; and double-shielded coaxial cables, where the 80 dB

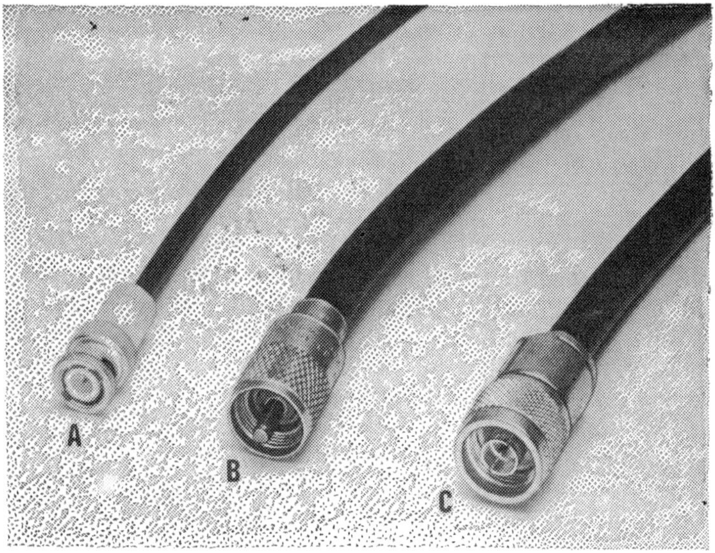

Fig. 6-11. The common coaxial connectors on their cables: (A) the BNC connector, (B) the UHF connector (formerly the PL-259), and (C) the type N connector.

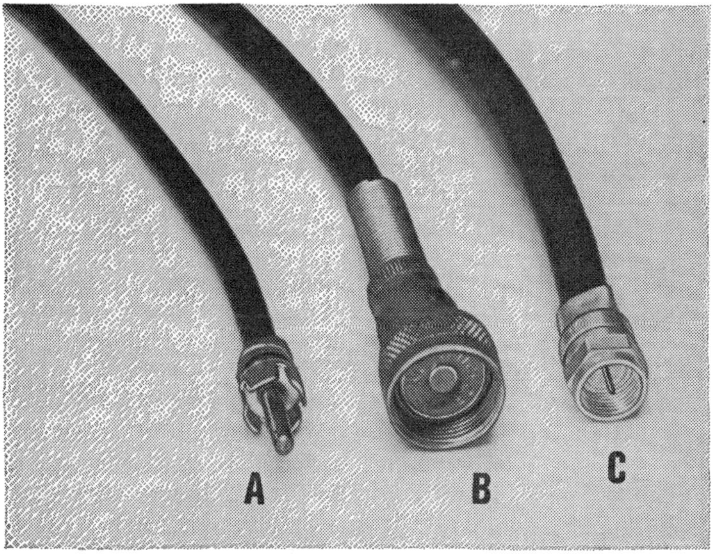

Fig. 6-12. The common shielded-cable connectors and the TV RF connector; (A) the phono connector, (B) the single-button microphone connector, and (C) the type F connector, used for television RF distribution.

shielding of a single braid is not sufficient and 120 dB shielding is required.

SPECIFICATIONS FOR PROBES

The low-capacitance or passive probe first gained its popularity with the oscilloscope. Its popularity in low-cost instrumention has only been recent, as the cost of this probe for what it could offer was rather high. With the price of this handy accessory substantially reduced and the sophistication of the measurements being made on the home experimenter's bench and service bench increasing all the time, the need for the passive probe has grown. It was recently noticed that the passive probe is a great adjunct to the digital frequency counter, as well as the oscilloscope. Both of these instruments have an input impedance of 1M resistance paralleled by 20–40 pF. Both instruments usually have BNC connectors for their inputs. Both are often involved in in-circuit, high-frequency measurements that will not tolerate the high capacitive loading of the simple shielded cable which has been used in the past.

Specifications for the passive probes include the multiplication factor, the circuit load capacitance and resistance the probe will show when working into an instrument with a specified input resistance and capacitance, and the range of input capacitance the instrument may have and still achieve compensation. The range of acceptable instrument input capacitance will usually lie between 20 and 40 pF. Other specifications include the maximum working voltage (peak AC plus DC), the cable length (probe input capacitance will vary with cable length), and the probe rise time.

Accessories available for this type of probe are important, as they add considerably to the ease of connection to the circuit under test. Probes having spring-loaded tips for grabbing circuit components should also have a replaceable tip, as these parts are often worn by use or damaged by contact with a hot soldering iron.

The passive probe is one of the few probes having any operating adjustments. A variable capacitor is usually provided so the probe may be adjusted (compensated) to make the capacitive ratio exactly equal to the resistive ratio. As the input capacitance of various instruments varies from one instrument to another (even between units of the same model), compensation must be made every time the probe is moved from one instrument to another, and often when moved from one input to another, and when moved from one input to another on the same instrument. Compensation of the probe is only important when one desires to use the probe in an application where amplitude measurements are being made.

To compensate the probe, it is connected to a square-wave source. A satisfactory source has a frequency about 1 kHz and a rise time of a few microseconds or better. The variable capacitor is adjusted until a square wave with good sharp edges is obtained without overshoot (Fig. 6-13).

The ×1 oscilloscope probe is used in measurements where input resistance and capacitance are not serious problems and where the full sensitivity of an instrument as well as the convenience of the probe mechanics are important. Usually, the ×1 probe is designed so the load capacitance it presents to the circuit under test is less than that presented by a simple shielded cable made with coaxial cable. Specifications for this probe will give the maximum operating voltage and the capacitance the probe will contribute to the circuit. The voltage rating of the probe will be given as DC plus peak AC. On the better units, the rise time of the probe will be specified with the probe connected to an oscilloscope of given input resistance and shunt capacitance. The series resistance of the probe center conductor may also be given.

Low-cost versions of these probes have been available for a number of years. Most of these probes offer minimal compensation, limited accuracy of attenuation, and most important, they generally do not offer any mechanical sophistication in making the connections to the circuit. It is this mechanical sophistication that is of partcular value to the user. With today's circuits employing many IC packages, tight foils, custom modules, and transistors in very small packages mounted close to the circuit board, any assistance in making mechanical connections to pins and foil is of great value.

There are some probes that offer the ×1 and ×10 feature in one unit. A switch selects between the two modes of operation. Generally speaking, these probes also fall in the classification of low-cost probes. Considering the investment in the oscilloscope or frequency counter, there is little justification in saving a few dollars on the probe. The frustration avoided by spending this $20 or so is well worth the money.

Radio-frequency probes usually have a specification indicating the frequency response or accuracy versus frequency. Accuracies are usually specified in decibels. Most probes have an upper limit from 100 to 300 MHz. The minimum input voltage level required to cause linear diode conduction is usually specified, starting near 0.2V–0.3V for the germanium or Schottky diode. Most diodes used in this service have a

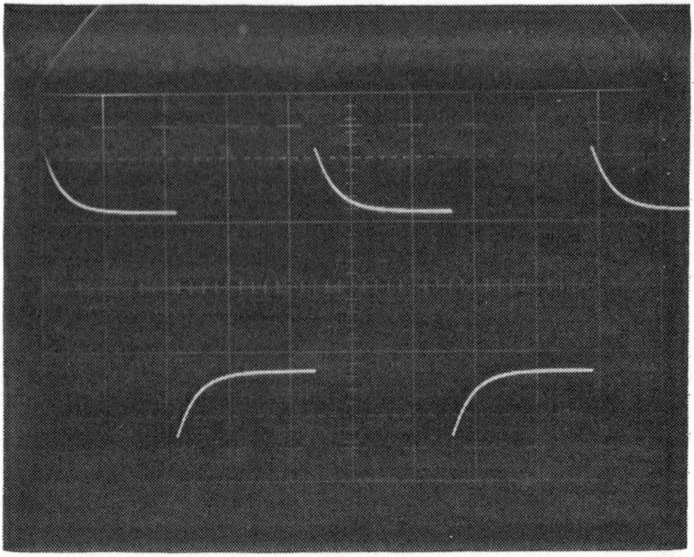

A

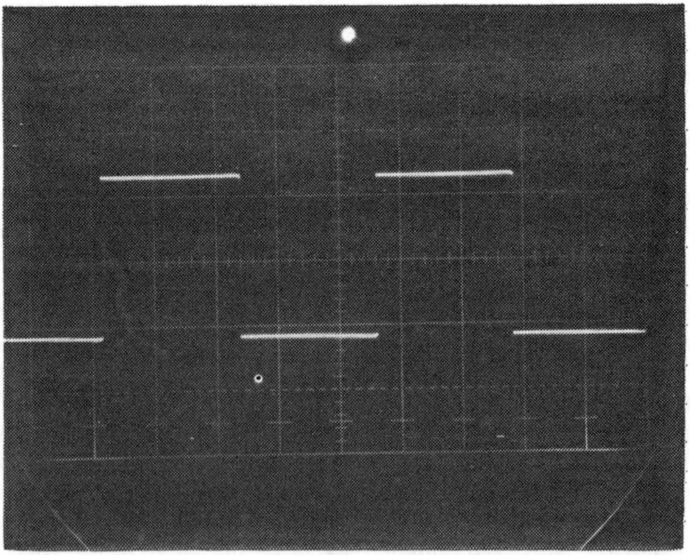

B

Fig. 6-13. Adjusting the passive probe for proper compensation. (A) Trimmer capacitance is set too low, and the capacitive division ratio of the probe is less than the resistive division ratio. Probe shows excessive

(continued on next page)

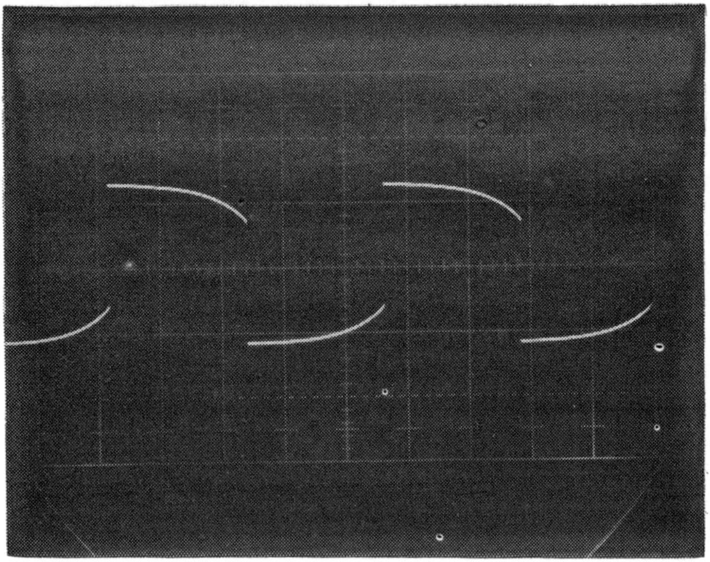

C

high-frequency response. (B) Proper compensation of the passive probe. The capacitive and resistive ratios are equal. (C) Trimmer capacitance is set too high and the capacitive division ratio of the probe is greater than the resistive division ratio. Probe shows poor high-frequency response.

reverse breakdown voltage around 30V. If greater potentials are to be measured, a number of diodes must be placed in series. A series of diodes cause the minimum voltage specification to increase by 0.3V or more per diode.

Demodulator probes fall into two categories: those designed for general-purpose applications and generally tailored for use with an oscilloscope, and those designed for use with some specific-purpose instrument. The general-purpose types are designed to operate into the 1M input impedance of an oscilloscope and provide demodulated bandwidths between 100 kHz and 1 MHz. Most of the basic specifications of the demodulator probe are similar to those of the RF probe. Demodulator probes are usually of the single-diode type.

Specific-purpose demodulator probes may either be part of an instrument, or they may be supplied as independent packages. The specifications applied to such demodulators will be determined by the intended application. Generally these probes will not suffice for universal service-oriented demodulator probes.

Manufacturers of high-voltage meter probes specify the multiplication factor, input impedance, accuracy of the multiplication, meter impedance the probe is designed to operate with, and maximum voltage the probe may be used to measure. The accuracy of these probes declines with humidity and temperature and with lack of cleanliness of the probe or multiplier resistor. The resistor must never be handled, as this creates a leakage path and lowers the accuracy of the voltage division. The HV probe must always be disconnected from the voltage source first. Disconnecting the probe from the meter first may damage the meter—and the operator.

ACCESSORIES

There are many adapters and other accessories to the probes and connectors which are useful. One of the most important is the 50-ohm feed-through terminator. This handy little device consists of a BNC male-to-female connector assembly shunted internally with a 50-ohm nonreactive resistance. The terminator allows the user to properly terminate a 50-ohm line at the 1M input to an oscilloscope; for example, when a non-terminated line would cause severe reflections and distort the measurement. Such terminators are invaluable for use on any pulse or RF system. Terminators usually come in 1W or 2W models. The voltage-standing-wave ratio (VSWR) is usually specified for these units; however, the importance of the VSWR specification is somewhat dubious, as the majority of the reflections that do occur are caused not by the terminator but by the capacitance of the device to which it is connected. Power levels of 1W or 2W will take care of most instrumentation uses. Remember, these are not kilowatt dummy loads for transmitters!

A 75-ohm version of the in-line terminator is available for video or TV RF work, and a 93-ohm terminator is also available.

Simple low-power attenuators are also avaiable in similar packages. These attenuators are available with a number of different attenuation levels and are capable of handling power in the 1W–2W area. These attenuators come in both 50- and 75-ohm models.

Other useful accessories inclued coaxial adapters, a number of which are shown in Fig. 6-14. The need for these devices will vary with the type of work being done. The various connections that can be made are self-explanatory. These devices will have some reflections, and the number should be held to a minimum wherever possible.

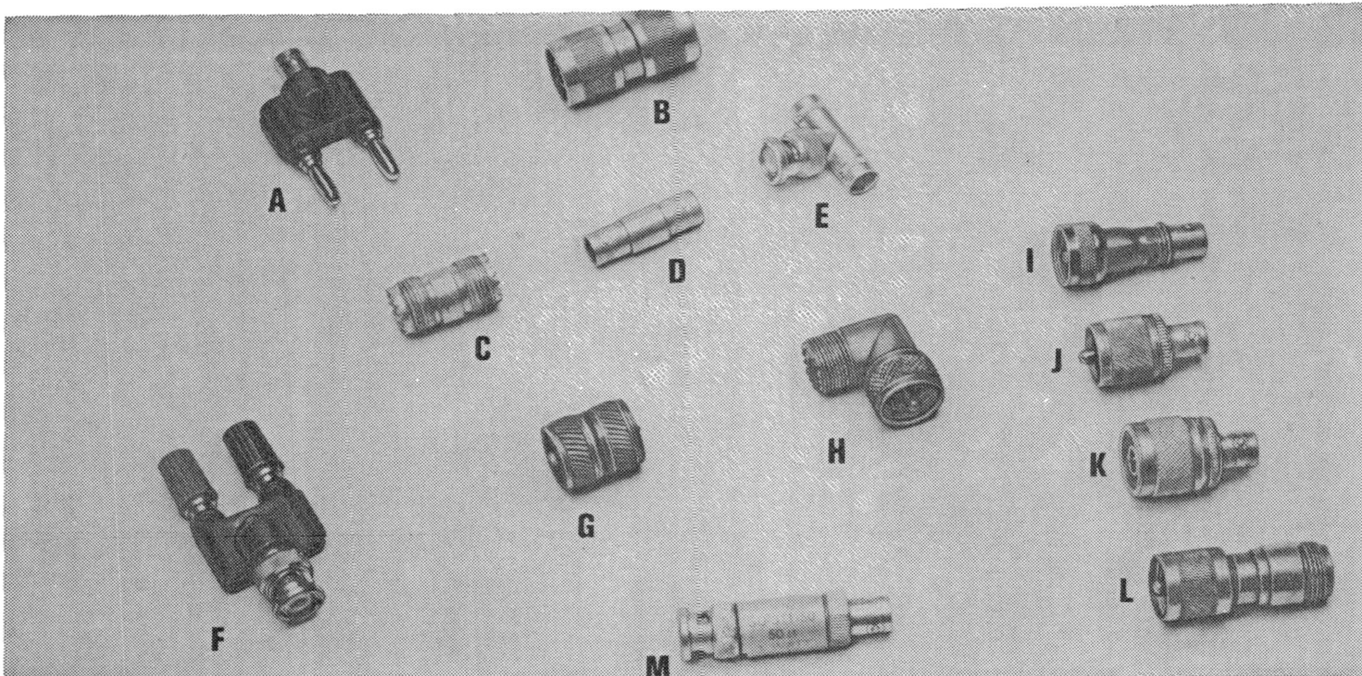

Fig. 6-14. A collection of common handy coaxial plug adapters. (A) BNC female to dual banana. (B) Type N male to type N male. (C) Female UHF to female UHF (PL-258). (D) Female BNC to female BNC. (E) BNC tee. (F) Dual ¾ in. 5-way binding posts to BNC male. (G) UHF male to UHF male. (H) UHF 90° elbow. (I) BNC female to single-button-microphone male (homemade). (J) BNC female to UHF male. (K) BNC female to type N male. (L) Type N female to UHF male. (M) A 1W, 50-ohm in-line terminator with BNC connectors.

SUMMARY

The investment in accessories, cables, and probes can be sizable. On the other hand, the time lost trying to make do when some simple device is not available can easily pay for the device. Of course the selection of accessories depends on the type of work being done. A number of cables will work as adapters. Such cables are often cheaper than adapters, and they can be made up as they are needed.

Probes, cables, and accessories can be the items that make an instrument what you thought it should be. A good selection of these items is well worth the comparatively small cost they represent. They are good items to have in mind when bargain hunting. Many of the adapters are available at the hamfest for instance, at a fraction of their original cost. A good collection of these devices can be quickly assembled, and a dot of enamel paint or the like enables quick identification should these devices become pooled with someone else's.

Appendix A
Decibel Ratios

A-1. Decibel Ratios

Change in Signal	Voltage or Current Change	Power Change
− 3 dB	X0.707	X0.5
− 6 dB	X0.5	X0.25
−10 dB	X0.333	X0.1
−20 dB	X0.1	X0.01
+ 3 dB	X1.414	X2
+ 6 dB	X2	X4
+10 dB	X3	X10
+20 dB	X10	X100

Appendix B
Characteristics of Commonly Used Coaxial Cables

B-1. Characteristics of Commonly Used Coaxial Cables

Type	Impedance, Ohms	Capacitance Per Foot	Voltage Rating	Center Conductor
RG-58/U	53.5	28.5	1900	20 Solid
RG-58A/U	50	29.5	1900	21 Stranded
RG-58C/U	50	29.5	1900	21 Stranded
RG-59/U	73	21.0	2300	22 Solid
RG-59A/U	75	20.5	2300	23 Stranded
RG-59B/U	75	20.5	2300	23 Stranded
RG-62/U	93	13.5	750	22 Solid
RG-62A/U	93	13.5	750	22 Solid

Appendix C
Specifications for Low-Cost Instruments

Specifications—Fluke 8000A DMM

DC VOLTAGE

Ranges	±199.9 mV, ±1.999 V, ±19.99 V, ±199.9 V, ±1199 V
Accuracy (1 year, 15°C to 35°C)	±(0.1% of reading + 1 digit)
Temperature Coefficient: (−10°C to 15°C and 35°C to 55°C)	±(0.01% reading/°C + .005% F.S/°C)
Input Impedance	10 MΩ, all ranges
Normal Mode Rejection	Greater than 60 dB @ 50 Hz, 60 Hz
Common Mode Rejection (1 kΩ unbalance)	Greater than 120 dB @ dc and 50 Hz, 60 Hz
Response Time	½ second
Maximum Input Voltage	1200 V rms, all ranges

AC VOLTAGE

Ranges	199.9 mV, 1.999 V, 19.99 V, 199.9 V, 1199V
Accuracy (1 year, 15°C to 35°C)	45 Hz to 10 kHz ±(0.5% of reading + 2 digits) 10 kHz to 20 kHz ±(1% of reading + 2 digits)
Temperature Coefficient: (−10°C to 15°C and 35°C to 55°C)	±(0.01% reading/°C +0.005% F.S./°C)
Input Impedance	10 MΩ in parallel with 100 pf
Common Mode Rejection (1 kΩ unbalance)	Greater than 60 dB @ 50 Hz, 60 Hz
Response Time	3 seconds, worst case
Maximum Input Voltage	1200 V rms, not to exceed 10^7 volts• Hz product on 20, 200, 1200 V ranges. 500 V rms on 200 mV and 2 V ranges.

DC CURRENT

Ranges	±199.9 μA, ±1.999 mA, ±19.99 mA, ±199.9 mA, ±1999 mA
Accuracy (1 year, 15°C to 35°C)	±(0.3% of reading + 1 digit)
Temperature Coefficient: (−10°C to 15° and 35° C to 55°C)	±(0.015% reading/°C +0.005% F.S./°C)
Voltage Burden	0.22V maximum up to 200 mA; 0.45V at 2 amps
Response Time	½ second
Maximum Input	2 amps rms (fuse protected)

AC CURRENT

Ranges	199.9 μA, 1.999 mA, 19.99 mA, 199.9 mA, 1999 mA
Accuracy (1 year, 15°C to 35°C)	45 Hz to 10 kHz ±(1% of reading + 2 digits) except 2000 mA range: 45 Hz to 3 kHz ± (1% of reading +2 digits)
Temperature Coefficient: (−10°C to 15°C and 35°C to 55°C)	±(0.015% reading/°C +0.005% F.S./°C)
Voltage Burden	0.22V maximum up to 200 mA; 0.45 V at 2 amps
Response Time	3 seconds, worst case
Maximum Input	2 amps rms (fuse protected)

RESISTANCE

Ranges	199.9Ω, 1.999 kΩ, 19.99 kΩ, 199.9 kΩ, 1999 kΩ, 19.99 MΩ
Accuracy (1 year, 15°C to 35°C)	±(0.2% of reading +1 digit) all ranges except: ±(0.5% of reading + 1 digit) on 20 MΩ range
Temperature Coefficient: (−10°C to 15°C and 35°C to 55°C)	KΩ ±(0.015% reading/°C +0.005% F.S./° C) 10 Meg ±(0.02% reading/°C +0.005% F.S./°C)
Response Time	½ second, all ranges except: 4 seconds on 20 MΩ range
Current through Unknown	200Ω range, 1 mA 2 kΩ range, 1 mA 20 kΩ range, 100 μA 200 kΩ range, 1 μA 2000 kΩ range, 1 μA 20 MΩ range, 0.1 μA
Maximum Input Voltage	200Ω and 2 kΩ ranges, 130 V rms 20 kΩ through 20 MΩ ranges, 250 V rms

GENERAL

Max. Common Mode Voltage	1200 V peak
Operating Temp. Range	−10°C to +55°C
Storage Temp. Range	−40°C to +75°C (−40°C to +60°C with batteries)
Humidity Range	0 to 80% RH

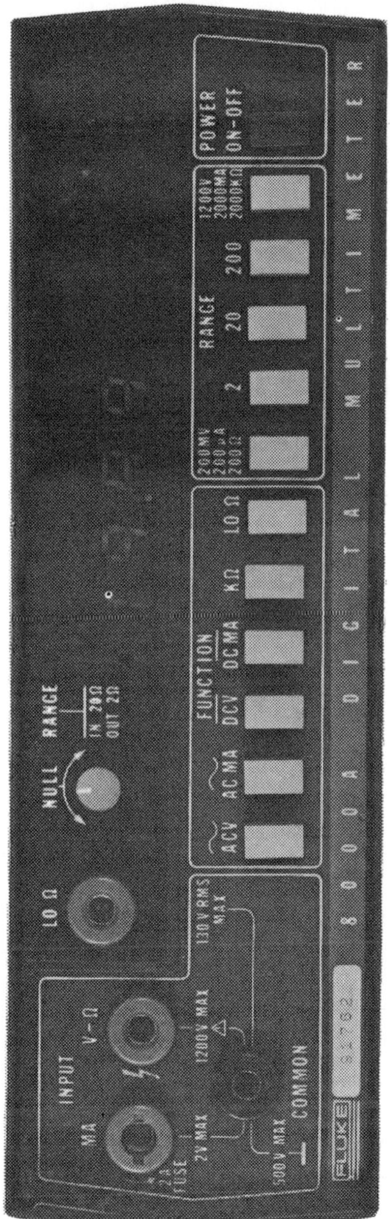

FLUKE 8000A DMM

Specifications—Heath IM-104 FETVOM

DC VOLTMETER

Nine Ranges . . . . 0–.1, .3, 1, 3, 10, 30, 100, 300, and 1000 volts full-scale.

Input Resistance . . . . . 10 MΩ.

Accuracy . . . . ±2% of full-scale (meter in horizontal position).

AC Rejection . . . . Accuracy not affected by impressed 60 Hz sine wave that has rms value two times greater than full-scale value.

Null Indicator . . . . . . On 0.1, 1, 10, 100, 1000 volt ranges, meter zero level may be adjusted to center of null reference scale. Null variations of approximately 1% of full-scale can be resolved.

AC VOLTMETER

Nine Ranges . . . . 0–.1, .3, 1, 3, 10, 30, 100, 300, and 1000 volts full-scale.

Range Factor . . . . 1 to 3.162 (10 dB) per step.

Input Resistance . . . . . 10 MΩ.

Input Capacity . . . . . . 100 pF typical.

Accuracy . . . . ±3% of full-scale. (Meter in horizontal position; 60 Hz sine wave.)

Response . . . . Responds to the average value of 1/2 of a sine wave. Scale calibrated to indicate rms.

Frequency Response . . .

Range	Error	Frequency
0.1 to 30	±3%	20 Hz to 20 kHz.
	±5%	20 Hz to 50 kHz.
100, 300	±3%	20 Hz to 10 kHz.
	±5%	20 Hz to 20 kHz.
1000	±3%	20 Hz to 1.0 kHz.
	±5%	20 Hz to 2.5 kHz.

DC MILLIAMMETER

Six Ranges . . . . 0–.01, .1, 1, 10, 100, 1000 milliamperes full-scale.

Voltage Drop (Approximate at full-scale) . . . . 0.100 volt (0.01 to 10 ranges).
0.150 volt (100 range).
0.350 volt (1000 range).

Insertion Resistance . . . . Voltage drop divided by full-scale current.

Accuracy . . . . . . ±2% on 0.01 to 100 ranges.
±3% on 1000 range.
(Meter in horizontal position.)

AC MILLIAMMETER

Six Ranges . . . . 0-0.01, .1, 1, 10, 100, 1000 milliamperes full-scale.

Voltage Drop (approximate at full-scale) . . . . 0.100 volt (0.01 to 10 ranges).
0.150 volt (100 range).
0.350 volt (1000 range).

Insertion Resistance . . . . Voltage drop divided by full-scale current.

Accuracy . . . . ±3% on 0.01 to 100 ranges.
±4% 1000 range.
(Meter in horizontal position; 60 Hz sine wave.)

Response . . . . Responds to the average value of 1/2 of a sine wave (DC current must be negligible). Scale-calibrated to indicate rms.

Frequency Response . . .

Range	Error	Frequency
0.01	±3%	20 Hz to 5 kHz.
0.1	±3%	20 Hz to 20 kHz.
1 to 1000	±3%	20 Hz to 50 kHz.

OHMMETER

Seven Ranges . . . . RX1 (10 Ω center scale), RX10, RX100, RX1K, RX10K, RX100K, RX1M.

Accuracy . . . . ±3 degrees of arc on Ω x 1 range (above 2 Ω.)
±2 degrees of arc on Ω x 10 to Ω x 1M ranges.
(Meter in horizontal position.)

Test Voltage (nominal) . . . 1.55 volts on +DC/Ω and –DC/Ω ranges.
0.085 volts on Ω_{LV} range.

dB RANGES

Nine Ranges
(0 dB = 1 mW into 600 Ω) . –40 to –18 (0.1 range).
–30 to –8 (0.3 range).
–20 to +2 (1 range).
–10 to +12 (3 range).
0 to +22 (10 range).
+10 to +32 (30 range).
+20 to +42 (100 range).
+30 to +52 (300 range).
+40 to +62 (1000 range).

Accuracy . . . . ±3 dB at –20 dB to ±0.3 dB at +2 dB.

HEATH IM-104 FETVOM

GENERAL

Voltage Dividers . . . . . 1% precision metal-film resistors.

Meter . . . . 4-1/2", 100 μA, 95 degree, ruggedized, taut-band movement.

Meter Protection . . . . . Will sustain a momentary overload of 220 volts AC or DC applied to any range.

2-ampere, quick-blow input fuse caused to open with overload on Ω x 1 range to Ω x 1 k range, and 0.01 to 1000 mA ranges by input clamping diodes.

FET protected on 0.1 volt to 1000 volt ranges and Ω x 10 k to Ω x 1 M ranges by clamping transistors.
Internal circuitry protected against improper battery installation.

	Meter movement protected by internal reverse-parallel bypass diodes.
Amplifier Circuit	Dual FET source followers driving operational amplifier.
Front Panel Switches	24-position, continuous-rotation RANGE switch [including battery check position BATT and OFF].
	4-position pushbutton switch: AC, Ω LV, –DC/Ω, +DC/Ω.
Meter Temperature Coefficient	Maximum of ±0.2% or ±0.2 degrees of arc (whichever applies) per degree C over a range of 15 to 35 degrees C.
Operating Temperature . . .	0 to 50 degrees C.
Storage Temperature (less batteries)	–40 to +80 degrees C.
Batteries Required	1-1.5 volt, D-cell (NEDA #13). 4-1.5 volt, AA-cell (NEDA #15).
Dimensions	4.95" wide x 6.95" high x 2.25" deep.
Weight	3 pounds (approximately), including batteries.

Specifications—Heath IM-2202 DMM

DC VOLTS

RANGE	INPUT IMPEDANCE	OVERRANGE CAPABILITY	OVERLOAD CAPABILITY
100.0 mV	50 MΩ	100%	300 VDC
1.000 V	500 MΩ	100%	300 VDC
10.00 V	10 MΩ	100%	1000 VDC
100.0 V	10 MΩ	100%	1000 VDC
1000 V	10 MΩ	None	1000 VDC

Accuracy . . . . Built-in References: ±0.5%, ±1 digit.
Laboratory Standards: ±0.2%, ±1 digit.

Temperature Coefficient . . . . ±200 PPM/°C.

AC VOLTS

RANGE	IMPEDANCE	OVERRANGE CAPABILITY	OVERLOAD CAPABILITY
100.0 mV	10 MΩ/120 pF	100%	250 VAC
1.000 V	10 MΩ/120 pF	100%	250 VAC
10.00 V	10 MΩ/60 pF	100%	750 VAC
100.0 V	10 MΩ/60 pF	100%	750 VAC
750 V	10 MΩ/60 pF	None	750 VAC

Accuracy . . . . Built-in References: ±1.5%, ±3 digits, from 50 Hz to 1 kHz (except 50 Hz to 10 kHz on 100 mV and 1 V ranges).

Laboratory Standards: ±.75%, ±3 digits, from 50 Hz to 10 kHz (except ±1.0%, ±3 digits, from 50 Hz to 2 kHz on 750 VAC range).

Temperature Coefficient . . . ±500 PPM/°C.

DC MILLIAMPERES

RANGE	VOLTAGE DROP (Approximate)	OVERRANGE CAPABILITY
100.0 μA	100 mV	100%
1.000 mA	100 mV	100%
10.00 mA	100 mV	100%
100.0 mA	150 mV	100%
1000 mA	300 mV	100%

Accuracy . . . . Built-in References: ±0.5%, ±1 digit (except ±1.0%, ±1 digit on 1000 mA range).

Laboratory Standards: ±0.2%, ±1 digit (except ±0.5%, ±1 digit on 1000 mA range).

Temperature Coefficient . ±250 PPM/°C.

Overload Protection . . . . Input coupled through 3.0 ampere fuse to shunt diodes for overloads above approximately 1.2 volts.

AC MILLIAMPERES

RANGE	VOLTAGE DROP (Approximate)	OVERRANGE CAPABILITY
100.0 μA	100 mV	100%
1.000 mA	100 mV	100%
10.00 mA	100 mV	100%
100.0 mA	150 mV	100%
1000 mA	300 mV	100%

Accuracy . . . . Built-in References: ±1.0%, ±3 digits from 50 Hz to 10 kHz (except ±1.5%, ±3 digits on 1000 mA range).

Laboratory Standards: ±0.5%, ±3 digits from 50 Hz to 10 kHz; (except ±1.0%, ±3 digits on 1000 mA range).

Temperature Coefficient . . ±550 PPM/°C.

Overload Protection . . . . . Input coupled through 3.0 ampere fuse to shunt diodes for overload above approximately 1.2 volts.

RESISTANCE

RANGE	TEST CURRENT	OVERRANGE CAPABILITY
100.0 Ω	1.0 mA	100%
1.000 kΩ	1.0 mA	100%
10.00 kΩ	0.1 mA	100%
100.0 kΩ	1.0 μA	100%
1000 kΩ	1.0 μA	100%

Accuracy	Built-in References: ±0.5%, ±1 digit. Laboratory Standards: ±0.2%, ±1 digit.
Temperature Coefficient . .	±300 PPM/°C.
Open-Circuit Voltage	12 VDC.
Overload Protection	Positive input overvoltage protection by blocking diode to 400 volts; negative input overvoltage through 1/32-ampere fuse and shunt diode.

GENERAL

Display	1999 Maximum; seven segment planar; 0.55" high digits.
Polarity Indication	Automatic "+" and "–".
Overrange Indication	Automatic above 1999; polarity sign and decimal point shown with flashing "1" in thousands.
Display Rate	Approximately 5 per second, nonblinking.
Normal Mode Rejection	35 dB (line operation); extended on battery operation.
Common Mode Rejection . . .	80 dB (line operation); extended on battery operation.
Power Line Isolation	May be operated 1500 V above power line ground (during line operation).
Operating Temperature Range .	10°C to 40°C (50°F to 104°F).
Storage Temperature Range . .	-40°C to 50°C (-40°F to 122°F) with batteries installed, -40°C to 70°C (158°F) with batteries removed.

HEATH IM-2202 DMM

Specifications—Heath SM-1212 DMM

Functions	DC volts, DC current, AC volts, AC current, ohms.
Ranges (Full Scale) .	DC volts: 0-2, 20, 200, 1,000 V. DC current: 0-2, 20, 200, 2,000 mA. AC volts: 0-2, 20, 200, 700 V rms (25 Hz to 10 kHz). AC current: 0-2, 20, 200, 2,000 mA rms (25 Hz to 10 kHz). Ohms: 0-200, 2K, 20K, 200K, 2,000K (2M) ohms.
Overrange	25% on all functions, within maximum input limits.
Maximum Input	3 amperes into AC or DC mA, and Ohms (fuse protected). 700 VAC rms into Volts (except 2 V range; 140 VAC rms). 1000 VDC into Volts (except 2 V range: 200 VDC).
Resolution (low range)	Volts: 10 mV. Current: 10 μA. Ohms: 1 ohm.
Display	2-1/2 digit numeric.
Accuracy (Full Scale ±1 digit)	DC volts ±1%. DC current ±1.5%. AC volts ±1.5%. AC current ±1.5%. Ohms ±2%.
Input Impedance . .	1 megohm on all voltage ranges. 2 V drop maximum on current ranges.
Sample Rate	Line frequency.
Power Requirements	110-130 VAC/220-260 VAC, 50/60 Hz, 8 watts.
Fuse Requirements .	Input: 3AG, 3 Ampere. Line: 3AG, 1/8 Ampere (1/16 Ampere for 240 VAC).
Dimensions (overall)	3-1/4" high, 9-1/2" wide, 8-3/4" deep. (8.25 cm high, 24 cm wide, 22.2 cm deep)
Weight	5 pounds (2.3 kg).

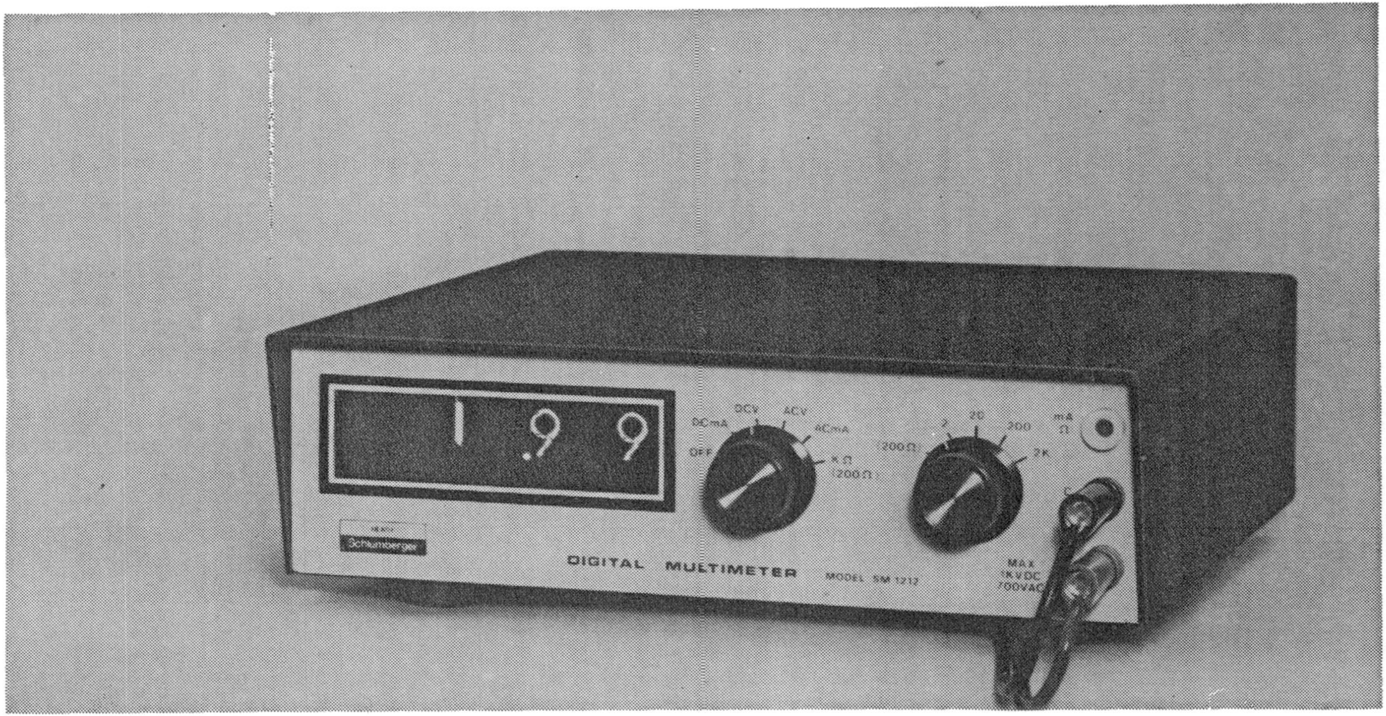

HEATH SM-1212 DMM

Specifications—Keithley 160B DMM

DC VOLTMETER

RANGE	MAXIMUM READING	ACCURACY ±(% of rdg + % of rng)		INPUT RESISTANCE	TEMPERATURE COEFFICIENT ±(% of rdg + % of rng)/°C		MAX. ALLOWABLE INPUT dc + peak ac
1 mV	1.999 mV	0.1%	0.1%	10 MΩ	0.01%	0.01%	600 V*
10 mV	19.99 mV	0.1%	0.1%	10 MΩ	0.01%	0.01%	600 V*
100 mV	199.9 mV	0.1%	0.1%	10 MΩ	0.01%	0.01%	600 V*
1 V	1.999 V	0.1%	0.1%	10 MΩ	0.01%	0.01%	1200 V
10 V	19.99 V	0.1%	0.1%	10 MΩ	0.01%	0.01%	1200 V
100 V	199.9 V	0.1%	0.1%	10 MΩ	0.01%	0.01%	1200 V
1000 V	1200 V	0.1%	0.1%	10 MΩ	0.01%	0.01%	1200 V

*1200 V momentary.

NORMAL MODE REJECTION RATIO: Greater than 80 dB above one digit on the 1-millivolt and 10-millivolt ranges, and greater than 60 dB above one digit on higher ranges, for voltages of line frequency or twice line frequency, with at least 1% of full rangeDC applied.

ZERO STABILITY: ±0.1 microvolts per °C.

COMMON MODE REJECTION RATIO (1 kΩ unbalance): Greater than 140 dB at dc and 40 to 200 Hz ac with at least 1% of full range dc applied. 1000 volts dc + peak-to-peak ACmaximum.

SETTLING TIME: Less than 2 seconds to within 0.1% of final reading, except less than 4 seconds on the 1 mV range.

DC AMMETER

RANGE	MAXIMUM READING	ACCURACY ±(% of rdg + % of rng)		SHUNT RESISTANCE	TEMPERATURE COEFFICIENT ±(% of rdg + % of rng)/°C		MAX. ALLOWABLE INPUT
10 nA	19.99 nA	0.3%	0.2%	1 MΩ	0.02%	0.02%	0.25 mA
100 nA	199.9 nA	0.2%	0.1%	100 kΩ	0.02%	0.01%	2.5 mA
1 μA	1.999 μA	0.2%	0.1%	10 kΩ	0.02%	0.01%	3 A*
10 μA	19.99 μA	0.2%	0.1%	1 kΩ	0.01%	0.01%	3 A*
100 μA	199.9 μA	0.2%	0.1%	100 Ω	0.01%	0.01%	3 A*
1 mA	1.999 mA	0.2%	0.1%	10 Ω	0.01%	0.01%	3 A*
10 mA	19.99 mA	0.2%	0.1%	1 Ω	0.01%	0.01%	3 A*
100 mA	199.9 mA	0.3%**	0.1%	0.1 Ω	0.01%	0.01%	3 A*
1000 mA	1999 mA	0.3%**	0.1%	0.1 Ω	0.01%	0.01%	3 A*

*Internally fused beyond 3 amperes.

**Self-heating due to long-term measurement of greater than 1 ampere can cause 0.1% additional error.

NORMAL MODE REJECTION RATIO: Greater than 60 dB above one digit for a current of line frequency or twice line frequency.

OHMMETER

RANGE	MAXIMUM READING	ACCURACY ±(% of rdg + % of rng)		VOLTAGE ACROSS UNKNOWN* at full range	TEMPERATURE COEFFICIENT ±(% of rdg + % of rng)/°C		MAX. ALLOWABLE INPUT rms ac+dc
1 Ω	1.999 Ω	0.4%	0.1%	1 mV	0.04%	0.03%	250 V
10 Ω	19.99 Ω	0.3%	0.1%	10 mV	0.04%	0.01%	250 V
100 Ω	199.9 Ω	0.3%	0.1%	100 mV	0.04%	0.01%	250 V
1 kΩ	1.999 kΩ	0.3%	0.1%	100 mV	0.04%	0.01%	250 V
10 kΩ	19.99 kΩ	0.3%	0.1%	100 mV	0.04%	0.01%	250 V
100 kΩ	199.9 kΩ	0.3%	0.1%	100 mV	0.04%	0.01%	250 V
1 MΩ	1.999 MΩ	0.3%	0.1%	100 mV	0.04%	0.01%	250 V
10 MΩ	19.99 MΩ	0.3%	0.1%	100 mV	0.04%	0.01%	250 V
100 MΩ	199.9 MΩ	10.0%	0.1%	100 mV	0.3 %	0.01%	250 V
1000 MΩ	1999 MΩ	30.0%	0.1%	100 mV	1.0 %	0.01%	250 V

*0.6 volt maximum into an open circuit.

GENERAL

DISPLAY: 3 digits plus 1 overrange digit; decimal position, polarity, and overload indication; 5 readings per second. 3 least significant digits blank when overload condition exists, except on 1000-volt range.

ISOLATION: Lo to power line ground: greater than 1000 megohms shunted by less than 300 picofarads. Lo may be floated up to ±1200 volts with respect to power line ground.

ANALOG OUTPUT: ±1 volt DCat up to 1 milliampere for full range input, 100% overranging except on the 1000-volt range, where overranging is 20%.

OFFSET CURRENT: Less than 10 picoamperes.

WARMUP: 1/2 hour to complete stabilization.

ENVIRONMENT:

Operating: 10°C to 35°C, 0% to 80% relative humidity. 0°C to 10°C and 35°C to 50°C, 0% to 80% relative humidity within twice computed accuracy.

Storage: -25°C to 70°C, without batteries installed.

POWER: 90-110, 105-125, 195-235 or 210-250 volts (switch selected), 50-60 Hz, 5 watts. Optional rechargeable 6-hour battery pack.

CONNECTORS: Input: binding posts. Analog Output: banana jacks.

DIMENSIONS, WEIGHT: Style K, 3-1/2 in. high x 9-1/4 in. wide x 10-3/4 in. deep (85 x 235 x 275 mm). Net weight, exclusive of batteries, 4 pounds (1,8 kg).

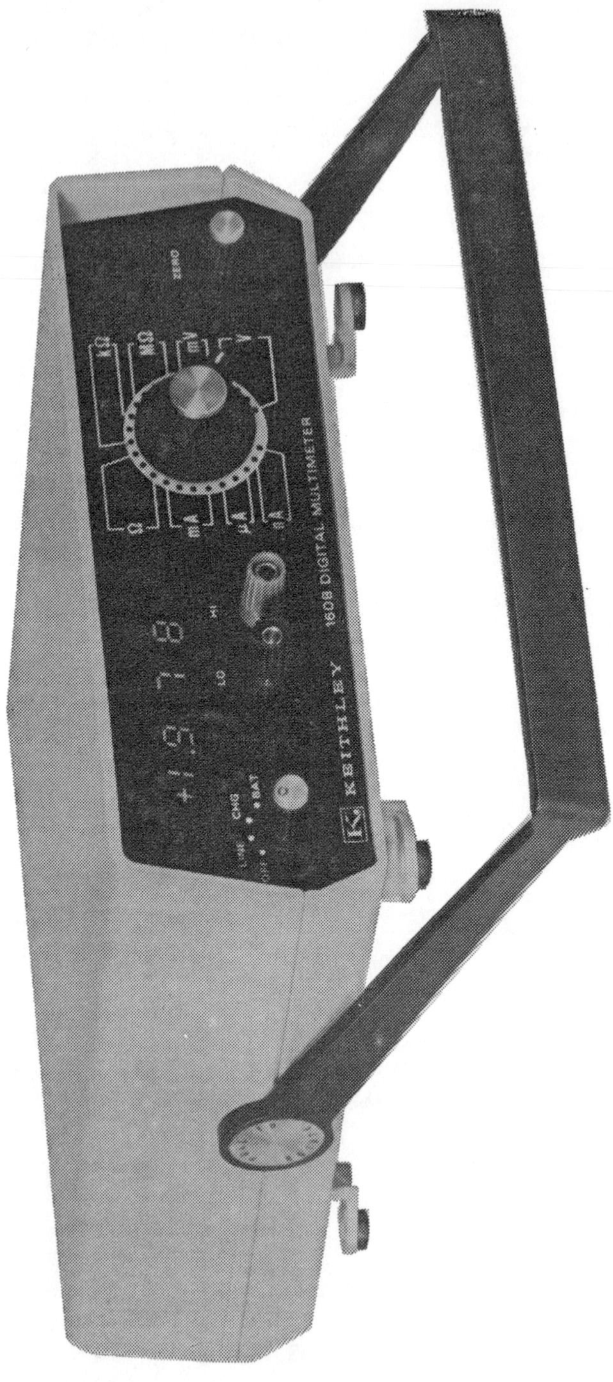

KEITHLY 160B DMM

Specifications—Keithley 168 DMM

DC VOLTMETER

RANGE	MAXIMUM READING	ACCURACY ±(% of rdg + % of rng)	
0.1V	.1999	0.1%	0.1%
1V	1.999	0.1%	0.1%
10V	19.99	0.1%	0.1%
100V	199.9	0.1%	0.1%
1000V	1000.	0.1%	0.1%

TEMPERATURE COEFFICIENT: ± (0.02% of reading + 0.01% of range) / °C.

INPUT RESISTANCE: 10 megohms.

NMRR: Greater than 75 dB over one digit 50 Hz to 20kHz, up to 300V p-p, with at least 1 mV dc applied.

CMRR (1kΩ unbalance): Greater than 140 dB at dc, 120 dB, 20 Hz to 20kHz (Lo driven).

AC VOLTMETER

RANGE	MAXIMUM READING	ACCURACY ±(% of rdg + % of rng)		FREQUENCY RANGE
0.1V	.1999	0.5%	0.3%	20Hz — 5kHz
1V	1.999	0.5%	0.3%	20Hz — 10kHz
10V	19.99	0.5%	0.3%	20Hz — 10kHz
100V	199.9	0.5%	0.3%	20Hz — 10kHz
1000V	500	2%	0.3%	20Hz — 5kHz

TEMPERATURE COEFFICIENT: ± (0.04% of reading + 0.01% of range) / °C.

INPUT IMPEDANCE: 9 megohms shunted by less than 90 picofarads.

CMRR (1kΩ unbalance): Greater than 100 dB dc to 65 Hz, 90 dB to 20kHz (Lo driven).

OHMMETER

RANGE	MAXIMUM READING	ACCURACY ±(% of rdg + % of rng) HI — mode — LO				VOLTAGE ACROSS UNKNOWN* @ 1100 digits HI — mode — LO		CURRENT IN UNKNOWN HI — mode — LO	
0.1kΩ	.1999	—	—	0.2%	0.2%	—	0.1V	—	1mA
1kΩ	1.999	0.2%	0.1%	0.2%	0.2%	1V	0.1V	1mA	100μA
10kΩ	19.99	0.2%	0.1%	0.2%	0.2%	1V	0.1V	100μA	10μA
100kΩ	199.9	0.2%	0.1%	0.2%	0.2%	1V	0.1V	10μA	1μA
1MΩ	1.999	0.2%	0.1%	0.2%	0.2%	1V	0.1V	1μA	0.1μA
10MΩ	19.99	0.2%	0.1%	—	—	1V	—	0.1μA	—

*6volts maximum in series with 9MΩ into an open circuit.

TEMPERATURE COEFFICIENT: ± (0.04% of reading + 0.01% of range) / °C.

AC AND DC AMMETER

RANGE	MAXIMUM READING	ACCURACY ±(% of rdg + % of rng) DC — mode — AC*				SHUNT RESISTANCE	FUSE PROTECTION
0.1mA	.1999	0.3%	0.1%	1%	0.3%	1.2kΩ	10mA
1mA	1.999	0.3%	0.1%	1%	0.3%	1.2kΩ	10mA
0.1 A	.1999	0.3%	0.1%	1%	0.3%	1.1 Ω	2 A
1 A	1.000	0.3%	0.1%	1%	0.3%	1.1 Ω	2 A

*30 Hz to 5 kHz.

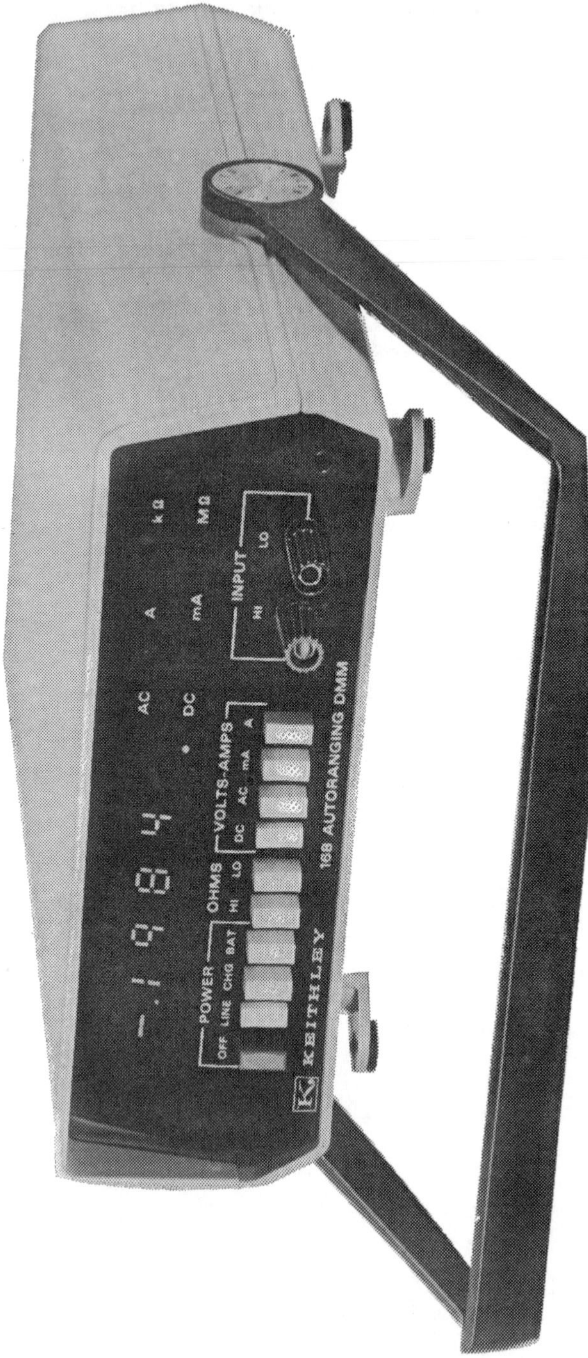

KEITHLY 168 DMM

TEMPERATURE COEFFICIENT:

DC ±(0.03% of reading + 0.01% of range)/ °C.
AC ±(0.05% of reading + 0.01% of range)/ °C.

GENERAL

ZERO STABILITY: ± 0.05% of range / °C (adjustable to zero with front panel control).

READING TIME: 3 seconds to within 0.1% of final reading including range changing.

DISPLAY: 3½ digits, appropriate decimal position, function and polarity indication. Upranges at 2000, downranges at 0189; five readings per second.

ISOLATION: Input Lo to power line ground, greater than 1000 megohms shunted by less than 300 picofarads. Maximum safe input between Lo and power line ground, 1200 volts peak.

POLARITY: Automatic.

RANGING: Automatic on each span.

OVERLOAD INDICATION: Display blinks when beyond specified maximum except on current ranges.

MAXIMUM ALLOWABLE INPUT: Electronically protected to ± 1200 volts (DC plus peak AC) on voltage ranges, 250V rms sine wave or DC on ohms. Fuse protected on current ranges.

ENVIRONMENT:
Operating: 0°C to 50°C.
0% to 70% relative humidity up to 35°C.
Storage: −25°C to +65°C.

POWER: 90-110, 105-125, 195-235 or 210-250 volts (switch selected), 50-60 Hz; 6 watts. Optional rechargeable 6-hour battery pack.

CONNECTORS: Binding Posts.

DIMENSIONS, WEIGHT: 3½ in. high x 9¼ in. wide x 10¾ in. deep (85 x 235 x 275 mm). Net weight, exclusive of batteries, 3½ pounds (1,6 kg).

Specifications—Systron-Donner 7004A DMM

DC VOLTS

Ranges: ±.10000V, ±1.0000V, ±10.000V, ±100.00V, ±1000.0V.

Resolution: ±10μV to ±0.1V in decade steps (dep. on range).

Overrange: 30% on all ranges except 1000V range. Maximum readings are ±.13000V, ±1.3000V, ±13.000V, ±130.00V and ±1000.0 V, at full accuracy.

Polarity Selection: Automatic with minus indicator.

Input Impedance: >1000 MΩ on .1V, 1V and 10V ranges; 10 MΩ on 100V and 1000V ranges.

Accuracy:

1, 10, 100, 1000 Volt Ranges (+25 ±5°C):
±0.01% rdg. ±0.01% f.s. for 1 month
±0.02% rdg. ±0.01% f.s. for 3 months

.1 Volt Range (+25 ±1°C):
±0.02% rdg. ±0.02% f.s. for 1 month

Temperature Stability (0 to 50°C):

1, 10, 100, 1000 Volt Ranges:
±0.001% rdg. ±0.001% f.s./°C

.1 Volt Range:
±0.005% rdg. ±0.01% f.s./°C

Full Scale Step Response: 1 sec. to rated accuracy.

Normal Mode Noise Rejection: $>$60 dB at 60 Hz.

Common Mode Noise Rejection: $>$120 dB at DC; $>$100 dB from 49 Hz to 1 kHz, with 1 kΩ source unbalance.

Maximum Input: ±1000 volts on any range without damage.

DC CURRENT

Ranges: ±.10000 mA, ±1.0000 mA, ±10.000 mA, ±100.00 mA, ±1000.0 mA.

Resolution: ±10 nA to ±0.1 mA in decade steps (dep. on range).

Overrange: 30% on all ranges. Maximum readings are ±.13000 n,A, ±1.3000 mA, ±13.000 mA, ±130.00 mA and ±1300.0 mA, at full accuracy.

Polarity Selection: Automatic with minus indicator.

Configuration: Shunts, internal to instrument.

Accuracy (for 3 months):
1, 10, 100, 1000 mA Ranges (+25 ±5°C): ±0.1% rdg. ±0.01% f.s.
.1 mA Range (+25 ±1°C): ±0.1% rdg. ±0.04% f.s.

Temperature Stability (0 to 50°C): ±0.005% rdg. ±0.002% f.s./°C.

Full Scale Step Response: 1 sec. to rated accuracy.

Normal Mode Noise Rejection: Same as DC VOLTS.

Common Mode Noise Rejection: Same as DC VOLTS.

Maximum Input: 100% above range selected without damage.

AC VOLTS

Ranges (rms value): .10000V, 1.0000V, 10.000V, 100.00V, 1000.0V.

Resolution: 10μV to 0.1 V in decade steps (dep. on range).

Overrange: 30% on all ranges except 1000 V range. Maximum readings are .13000V, 1.3000V, 13.000V, 130.00V, 500.00V, rms.

Input Impedance: 1 MΩ shunted by 100 pF.

Accuracy (for 3 months), 50 Hz to 20 kHz:
1, 10, 100, 1000 Volt Ranges (+25 ±5°C): ±0.2% rdg. ±0.02% f.s.
.1 Volt Range (+25 ±1°C): ±0.5% rdg. ±0.2% f.s.

Full Scale Step Response: 3 sec. to rated accuracy.

Maximum Input: 500 volts rms on 10, 100 and 1000 volt ranges; 150 volts rms on 0.1 and 1 volt ranges, without damage.

AC CURRENT

Ranges: .10000 mA, 1.0000 mA, 10.000 mA, 100.00 mA, 1000.0 mA, rms.

Resolution: 10 nA to 0.1 mA in decade steps (dep. on range).

Overrange: 30% on all ranges. Maximum readings are .13000 mA, 1.3000 mA, 13.000 mA, 130.00 mA and 1300.0 mA.

Configuration: Shunts internal to instrument.

Accuracy (for 3 months), 50 Hz to 20 kHz:
1, 10, 100, 1000 mA Ranges (+25 ±5°C): ±0.3% rdg. ±0.02% f.s.
.1 mA Range (+25 ±1°C): ±0.5% rdg. ±0.2% f.s.

Full Scale Step Response: 3 sec. to rated accuracy.

Maximum Input: 100% above range selected without damage.

RESISTANCE

Range	Current thru Rx	Resolution
1.0000 kΩ	5 mA	0.1 ohm
10.000 kΩ	0.5 mA	1 ohm
100.00 kΩ	50 μA	10 ohms
1000.0 kΩ	5 μA	100 ohms
10.000 MΩ	0.5 μA	1000 ohms

Overrange: 30% on all ranges. Maximum readings are 1.3000 kΩ, 13.000 kΩ, 130.00 kΩ, 1300.0 kΩ and 13.000 MΩ at full accuracy.

Configuration: Two-wire measurement system.

Accuracy (+25 ±5°C): ±0.1% rdg. ±0.01% f.s. on all ranges.

Full Scale Step Response: Typ. 1 sec. on all kΩ ranges and 3 sec. on 10 MΩ range.

Voltage Protection: 130 Vrms without damage on any range.

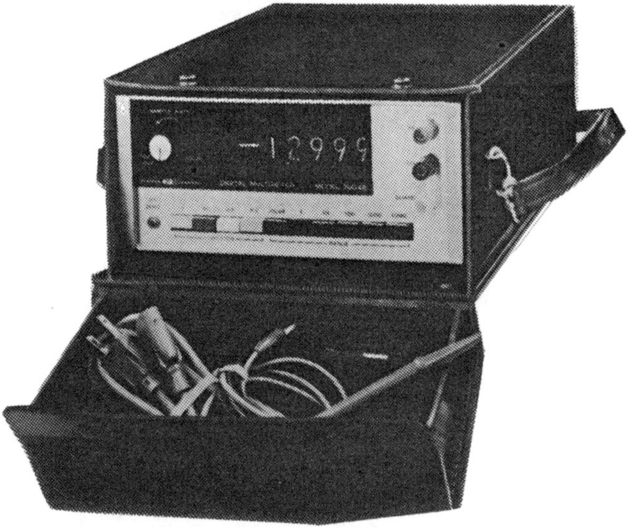

SYSTRON-DONNER 7004A DMM

Specifications—Fluke 1941A Counter

FREQUENCY MEASUREMENTS

Range	5 Hz to 40 MHz (filter out, 5 Hz to 1 MHz (filter in)
Gate Times	1 msec, 10 msec, 100 msec, 600 msec, 1 sec
Resolution	1 Hz at 1 sec gate time decreasing to 1 kHz at 1 msec gate time
Accuracy	±1 count ± time base accuracy
TOTALIZING	Counts to 999.999, then activates overflow indicator

RPM — Determined by pulses per revolution from source. 5 pulses/sec. to 40 million pulses/ sec. limits, with 999.999 display capacity.

Decimal placed for 100 pulses per revolution tachometer.

INPUT SENSITIVITY

Sine Wave — 40 mV rms (attenuator out)
400 mV rms (attenuator in)

Pulse — 150 mV p-p (attenuator out) at minimum pulse width of 15 nsec.
1.5V p-p (attenuator in)

Impedance — 1 MΩ in parallel with 30 pf (attenuator and filter out)

Attenuator — Sensitivity is decreased by a factor of 10 in the X10 position.

Filter — 1 MHz low pass filter activated by front panel push button.

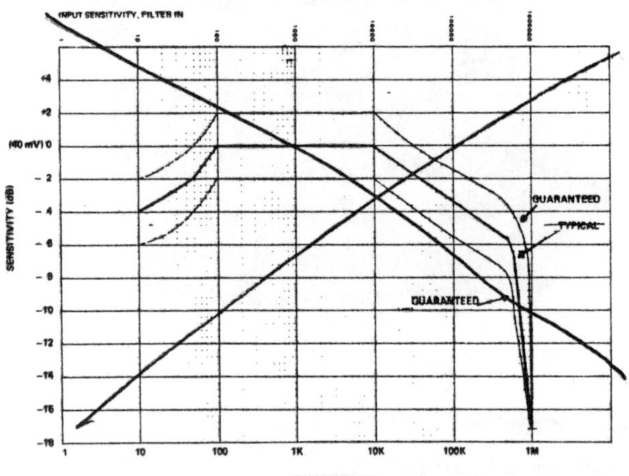

Note: 1941A must be terminated with 50 ohms.

Overload Capability — DC +Peak AC must not exceed 250V

150V rms to 1 kHz decreasing to 5V rms at 40 MHz Attenuator Out and 50V rms at 40 MHz Attenuator in.

TIME BASE

Stability

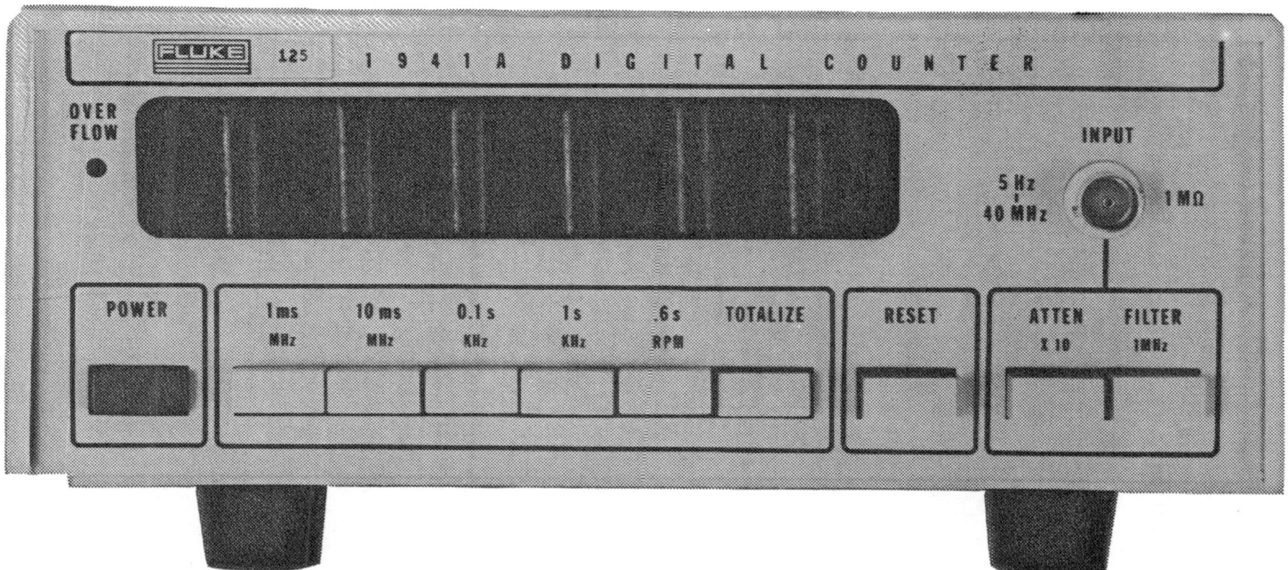

FLUKE 1941A COUNTER

Aging Rate	< 2 parts in 10^6/ms
Temperature	< ± 5 parts in 10^6, 0° to 50°C
Typically	< ± 2 parts in 10^6, 20° to 30°C
Line Voltage	< ± 2 parts in 10^7 for 10% line variation

External Time base Input

Frequency	10 MHz
Sensitivity	500 mV rms
Impedance	20 kΩ shunted with 50 pF

BCD OUTPUT OPTION—02

Data	4 lines per digit: 8421 Parallel BCD Logic 1 is true Overflow flag: Logic 1 is true Print command: Logic 1 is true Pulse width 100 μS Delayed Print command: Logic 0 is true Pulse width 100 μS Delay Time 200 μS
Logic Levels	DTL/TTL compatible positive logic Logic 1 = +3.5 volts minimum Logic 0 = +0.8 volts maximum
Loading	Logic will drive eight 54/74 series TTL loads

BCD OFFSET OPTION—05

Input	4 lines per digit, 8421 Parallel BCD Logic 1 is true
Output	Overflow flag - Logic 1 is true
Logic Levels:	DTL/TTL Compatible positive logic or contact closure Logic 1 = +3.5 volts minimum or open circuit Logic 0 = +0.8 volts maximum or ground
Loading	One 54/74 series TTL load per input line

Specifications—Heath IB-1100 Counter

Frequency Range	1 Hz to over 30 MHz.
Accuracy	±1 digit ± time base stability.
Sensitivity (after 30-minute warmup)	100 mV rms typical, 250 mV rms maximum.
Gate Time	1 millisecond or 1 second; with automatic reset.
Input Impedance	1 MΩ, shunted by less than 20 pF.
Maximum Input Voltage	150 V rms to 100 kHz. Derate 48 volts per decade above 100 kHz. (Maximum DC input is 200 volts.)

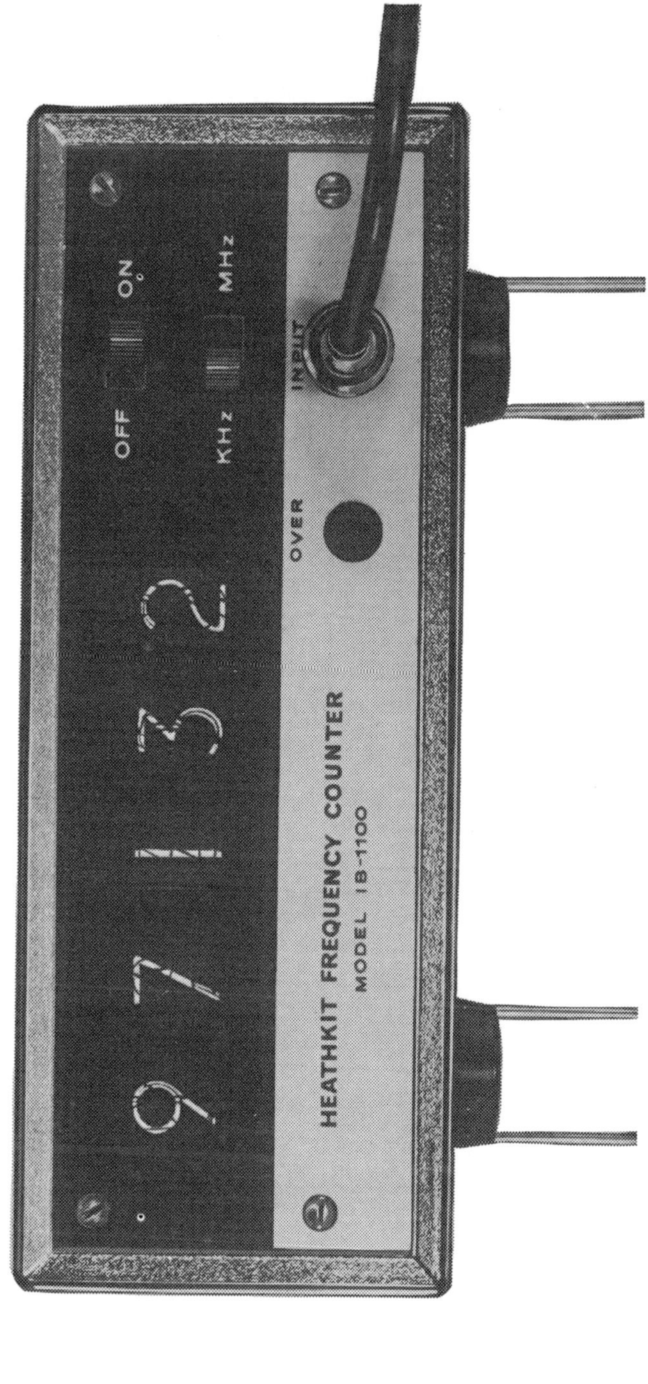

HEATH IB-1100 COUNTER

Time Base Frequency . . . . 1 MHz.

Time Base Stability (After 30-minute warmup) . <±3 PPM between 22°C and 37°C.
<1 PPM/Month after 30 days operation.
<±20 PPM between 10°C and 40°C.

Front Panel Indications . . . . Five 10-numeral, cold-cathode tubes, plus a neon lamp for Overrange indication.

Rear Panel . . . . Time Base Osc adjustment, TP Level control, TP Output terminal.

Front Panel . . . . Power OFF-ON switch, Range MHz/kHz switch, and BNC Input connector.

Ambient Temperature Range . . . . Storage – 55°C to 80°C.
Operating 10°C to 40°C.

Power Requirements . . . . 110-130 or 220-260 Vac, 50/60 Hz, 15 watts.

Dimensions . . . . 7-1/4" wide x 3-1/8" high x 9-1/4" deep.

Net Weight . . . . 4 pounds.

Specifications—Heath SM-110C Counter

INPUTS

Frequency Range:

A Input . . . . 1 Hz to 65 MHz minimum.

B Input . . . . 10 MHz to 200 MHz minimum.

Prescale (SM-110C only) . . . . 40 MHz to 600 MHz minimum.

Sensitivity:

A Input . . . . 10 mV rms.

B Input . . . . 15 mV rms.

Prescale (SM-110C only) . . . . 20 mV rms.

Input Impedance:

A Input . . . . 1 MΩ shunted by 15 pF.

B Input . . . . 50 Ω with ≤ 2 V rms input.

Prescale (SM-110C only) . . . . 50 Ω with ≤ 2 V rms input.

Maximum Input:

A Input . . . . See AC Derating Curve, Figure 1-1. (Maximum DC input is 250 V.)

B Input . . . . 5 V rms.

TIME BASE (Referenced to 24°C after 1/2-hour warmup.)

SM-110A:

Frequency . . . . 1 MHz.

Setability . . . . ±0.1 ppm.

Maximum Aging Rate . . . 7.5 ppm/yr (1 ppm/mo).

Temperature Stability . . . . ±10 ppm, maximum, 0° to 40° C ambient.
Frequency . . . . 1 MHz.

Setability . . . . ±0.1 ppm.

Maximum Aging Rate . . . . . ±1 ppm/yr (3 x 10^{-7}/mo).

Short Term Stability . . . . . ±1 x 10^{-9}/0.25 sec.

Temperature Stability . . . . ±1 ppm maximum, 0° to 40° C ambient.

Line Voltage Stability . . . . ±0.5 ppm, 105-130 Vac.

External Oscillator:

Frequency . . . . DC to 2 MHz.

Sensitivity . . . . TTL or 2.5 V rms at Rs 50 Ω.

Protection . . . . –5 V peak to +10 V peak.

RANGES—GATE INTERVAL

SM-110A, B, and C . . . . Hz – 10 sec, kHz – 1 sec, MHz – 0.1 sec, MHz – 0.01 sec.

Remote (SM-110B and C) . . . . . . Hz – 10 sec, kHz – 1 sec, MHz – 0.1 sec, MHz – 0.01 sec, MHz – 0.001 sec, GHZ – 100μsec, GHz – 1 μsec.

GENERAL

Display Time . . . . 0.2 sec plus Gate Interval.

Remote Programming
(SM-110B and C)

Inputs . . . . TTL level data required to control Range, Input Select, Reset, and Count Inhibit.
Logic 0≤0.8V. Logic 1≥2.0V but >5.5V.

Outputs . . . . 7 digits of BCD (8-4-2-1), Overrange flag, decimal points, print command, 5 V reference (will drive all remote programming) and circuit ground.

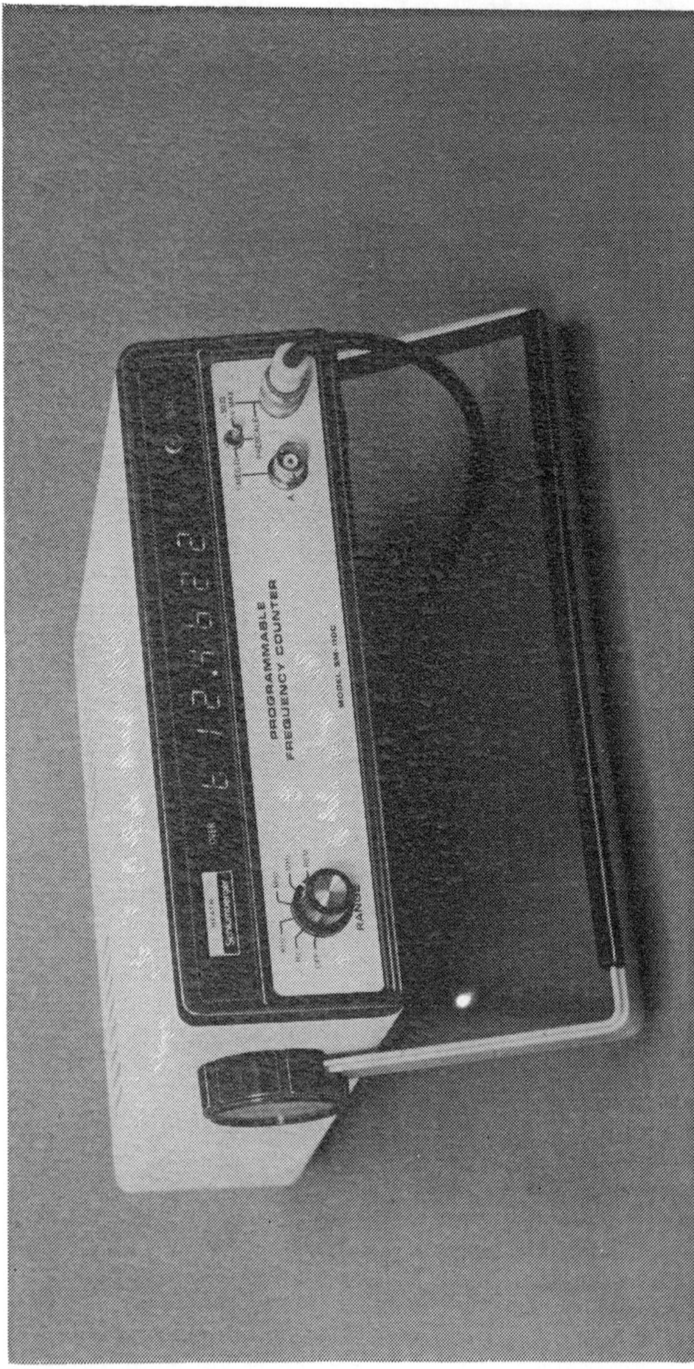

HEATH SM-110C COUNTER

Mating Connector. . (not supplied)	Cinch DB-25P or equivalent (2 required).
Power Requirements . . .	105-130 V, 50/60 Hz, 45 watts maximum. (210-260 V by changing internal switch and fuse.)
Fuse:	
125 V Operation . .	1/2-ampere, 125 V slow-blow.
250 V Operation . .	1/4-ampere, 250 V slow-blow.
Dimensions	12-7/16" deep, 8-1/2" wide, 2-3/4" high.
Net Weight	7-1/2 pounds.

Specifications—Systron-Donner 6202A Counter

FREQUENCY MEASUREMENT

Range: 20 Hz to 5 MHz

Sensitivity: 25 mV rms sine wave from 20 Hz to 1 MHz; 50 mV rms from 1 MHz to 5 MHz. 150 mV pk-pk at pulse width of 100 ns; 300 mV pk-pk at min. pulse width of 50 ns.

Resolution: 0.1 Hz to 100 Hz in decade steps

Accuracy: ±1 count ± time base accuracy

Reads In: kHz with autopositioned decimal point

TOTALIZE MEASUREMENT

Rate Limit: 20 Hz to 5 MHz

Register Capacity: 10^5 events (10^7 events with option 16)

Control: Manual front panel pushbutton for Totalize/Hold

INPUT CHARACTERISTICS

Input Impedance: 1 MΩ shunted by 45 pF (dynamic)

Coupling: AC

Maximum Input (without damage): 500 volts peak (DC plus AC components). 250 V rms, 20 Hz to 10 kHz; 50 V rms, 10 kHz to 1 MHz; 10 V rms, 1 MHz to 5 MHz.

GENERAL

Display: 5 digits standard (7 digits optional)

Display Storage: Provides continuous display of last reading

Temperature Range: 0°C to 50°C

Dimensions: 3¼"H x 7"W x 8-1/8"D (8.2 x 17.8 x 20.6 cm)

Weight: 5 lbs (2.3 kg) net; 7½ lbs (3.4 kg) shipping

Power: 100/115 or 200/230 volts (±10%), 50 to 400 Hz. 12 watts.

Price: $435

SYSTRON-DONNER 6202A COUNTER

Specifications—Systron-Donner 6220A, 6251, 6252 Counters

FREQUENCY MEASUREMENTS

Frequency Ranges

"A" Input: 20 Hz to 50 MHz (all models)

"B" Input: 25 MHz to 180 MHz (Model 6251); 25 MHz to 512 MHz (Model 6252)

Tone Ranges (via "A" Input)

50 Hz to 1 kHz (0.001 Hz resolution in one second.

1 kHz to 10 kHz (0.01 Hz resolution in one second.

Accuracy: ±1 count ± time base accuracy

Reads In: Hz or kHz with autopositioned decimal point

RATIO MEASUREMENTS

Display: F_A/F_{EXT}. Time multiplier from 10^3 to 10^6 in decade steps

Ranges

F_A Input: 20 Hz to 50 MHz (via "A" input)

F_{EXT} Input: 1 kHz to 10 MHz (via External Time Base Input)

INPUT CHARACTERISTICS

Input Impedance and Coupling

"A" Input: 1 MΩ shunted by 25 pF. AC coupled

"B" Input: 50 ohms nominal. AC coupled

Sensitivity

"A" Input: 25 mV rms

"B" Input: 50 mV rms

Attenuation: Variable sensitivity control on "A" input

Maximum Input

"A" Input: The following voltage levels may be applied without damage: 500 volts peak (DC plus AC component); 250 volts rms from 20 Hz to 10 kHz; 50 volts rms from 10 kHz to 2 MHz; 10 volts rms from 2 MHz to 50 MHz.

"B" Input: +20 dBm with overload protection. Levels in excess of +20 dBm will open relay approximately 20 mS after overload is applied. Front panel pushbutton to reset overload; Panel meter indicates correct level range for input signal.

STANDARD TIME BASE

Crystal Frequency: 10 MHz

External Time Base Input: Accepts 1 kHz to

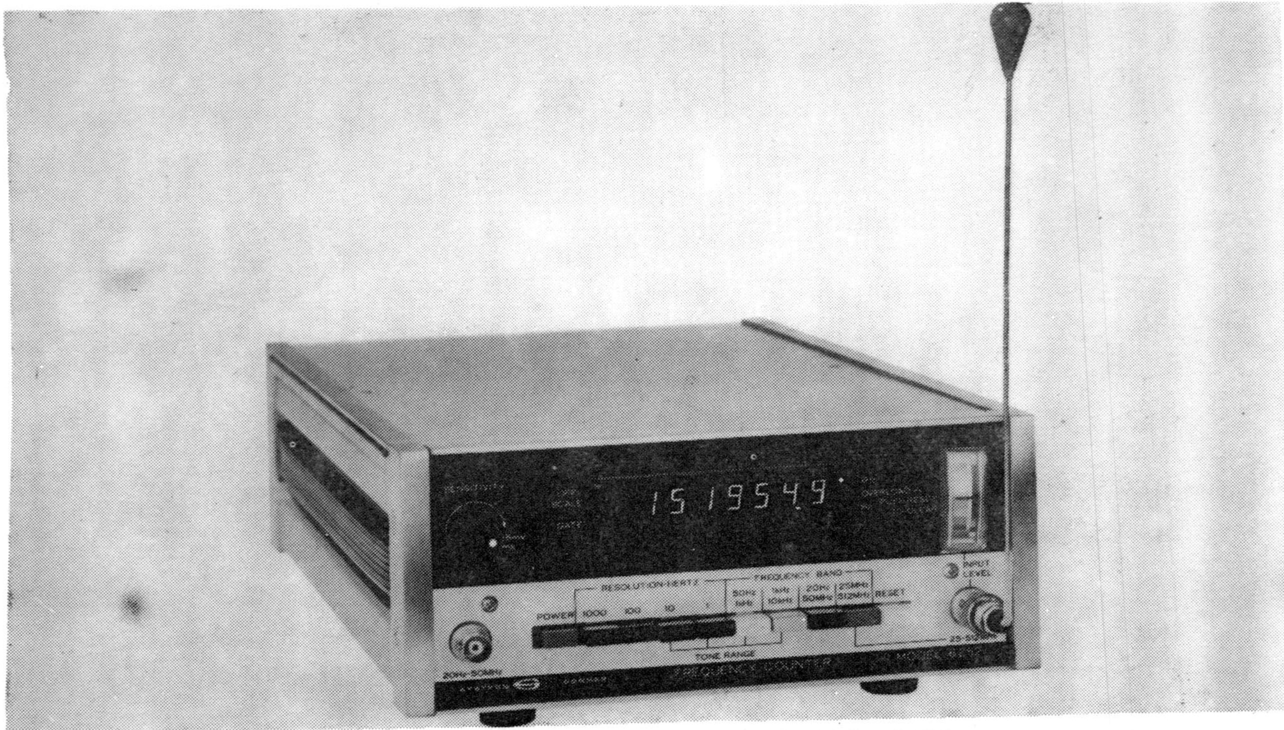

SYSTRON-DONNER 6220A, 6251, 6252 COUNTERS

10 MHz input with sensitivity of 1 volt rms and maximum input of 10 volts peak-to-peak. Input impedance is 500 ohms

GENERAL

Display: 8-digit in-line readout with off-scale indicator for overflow. Automatic leading-zero suppression. Display storage holds reading between samples.

Operating Temperature: 0°C to +50°C

Power: 100/115 or 200/230 volts (±10%), 48 to 440 Hz. Approx. 20 watts

Dimensions: 3-1/2"H x 8-3/8"W x 13-1/2" D (8.9 x 21.3 x 34.2 cm)

Weight: 10 lbs. (4.5 kg) net; 17 lbs. (7.7 kg) shipping.

Price:

Model 6220A:	$ 650.00
Model 6251:	985.00
Model 6252:	1195.00

Specifications—Systron-Donner 6250A Counter

FREQUENCY MEASUREMENT

Range: 20 Hz to 50 MHz

Resolution: Automatic, 1 Hz; Manual, 0.1 to 100 Hz selectable.

Display: kHz, with autopositioned decimal point.

Accuracy: ±1 count ± time base accuracy

PERIOD MEASUREMENT

Range: 1 μs to 50,000 μs with sinewave input; 1 μs to 999999.99 μs with pulse or square wave input.

Resolution

Automatic: 0.01 μs to 0.0000001 μs by period averaging 100 to 10^6 periods. Automatically searches multiple average required to fill the register.

Manual: Selection of multiple period average from 10 to 10^6 periods. Counts 10 MHz.

Display: μsec, with autopositioned decimal point

Accuracy: ±1 count ± trigger error ± time base accuracy

TIME INTERVAL (A –) MEASUREMENT

Range: 0.001 ms to 9999999.9 ms

Resolution: 0.0001 ms to 0.1 ms

Slope Selection: Independent positive or negative for Start and Stop inputs.

Display: msec, with autopositioned decimal point.

RATIO MEASUREMENT

Function: f_A / f_B x multiplier (M)

Frequency Range: f_A input, 20 Hz to 50 MHz; f_B input, 20 Hz to 1 MHz.

Multiplier (M) Range

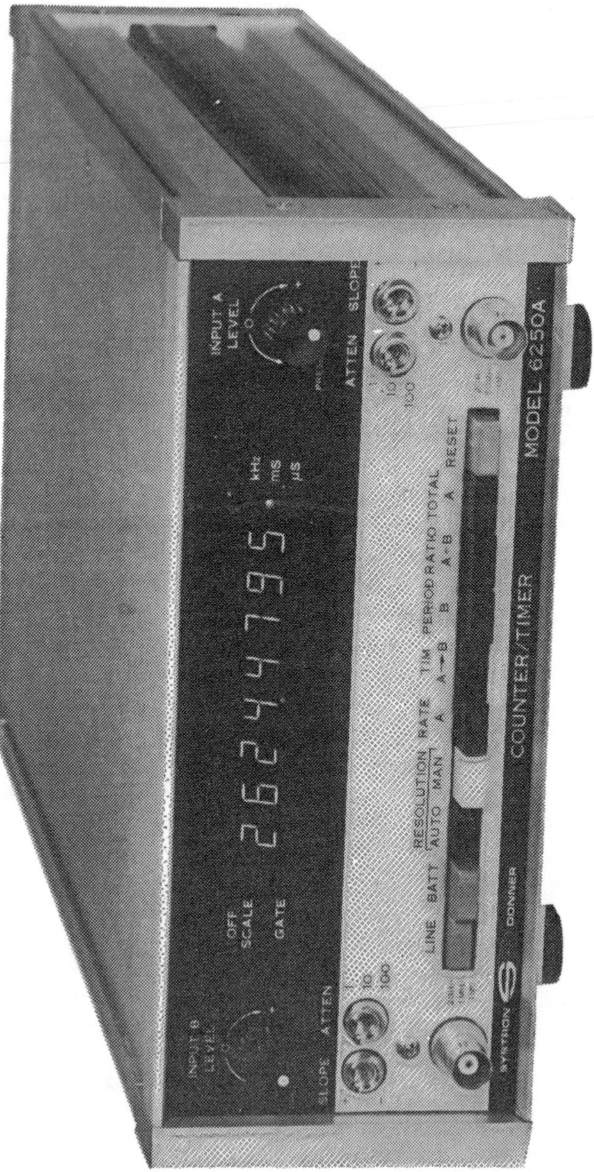

SYSTRON-DONNER 6250A COUNTER

Manual: M = 10 to 10^6
Automatic: M = 100 to 10^6. Automatically searches multiple average required to fill the register.

TOTALIZE

Rate Limit: 20 Hz to 50 MHz
Register Capacity: 10^8 events
Control: Front panel pushbutton for Totalize/Hold

Specifications—Heath IO-102 Oscilloscope

VERTICAL CHANNEL

Input Impedance . . . . 1 megohm shunted by 35 pF.

Sensitivity . . . . 30 mV peak-to-peak/centimeter (uncalibrated).

Maximum Input Voltage . . . . 600 V peak-to-peak.

Frequency Response . . . . DC to 5 MHz ±3 dB.

Rise Time . . . . 80 nanosecond.

Attenuator . . . . 3 position, compensated, X1, X10, X100.

HORIZONTAL CHANNEL

Input Impedance . . . . 1 megohm shunted by 50 pF.

Sensitivity . . . . 30 mV peak-to-peak/centimeter (uncalibrated)

Maximum Input Voltage . . . . . 600 V peak-to-peak.

Frequency Response . . . . DC to 5 MHz ±3 dB.

SWEEP GENERATOR

Type . . . . Recurrent, automatic sync.

Range . . . . 10 Hz to 500 kHz in five ranges.
CRT . . . . 5DEP1, 6 x 10 centimeter viewing area green, medium persistence phosphor.

Power Supplies . . . . All solid-state rectifiers.
All amplifier supplies regulated.

Power Requirements . . . . 110-130 or 220-260 Vac, 50/60 Hz, 35 watts.

Overall Dimensions . . . . 12-3/4" high, 9-1/4" wide, 16-1/4" long. These dimensions include all protruding surfaces; knobs, handle, feet, etc.
Net Weight . . . . 27 lbs.

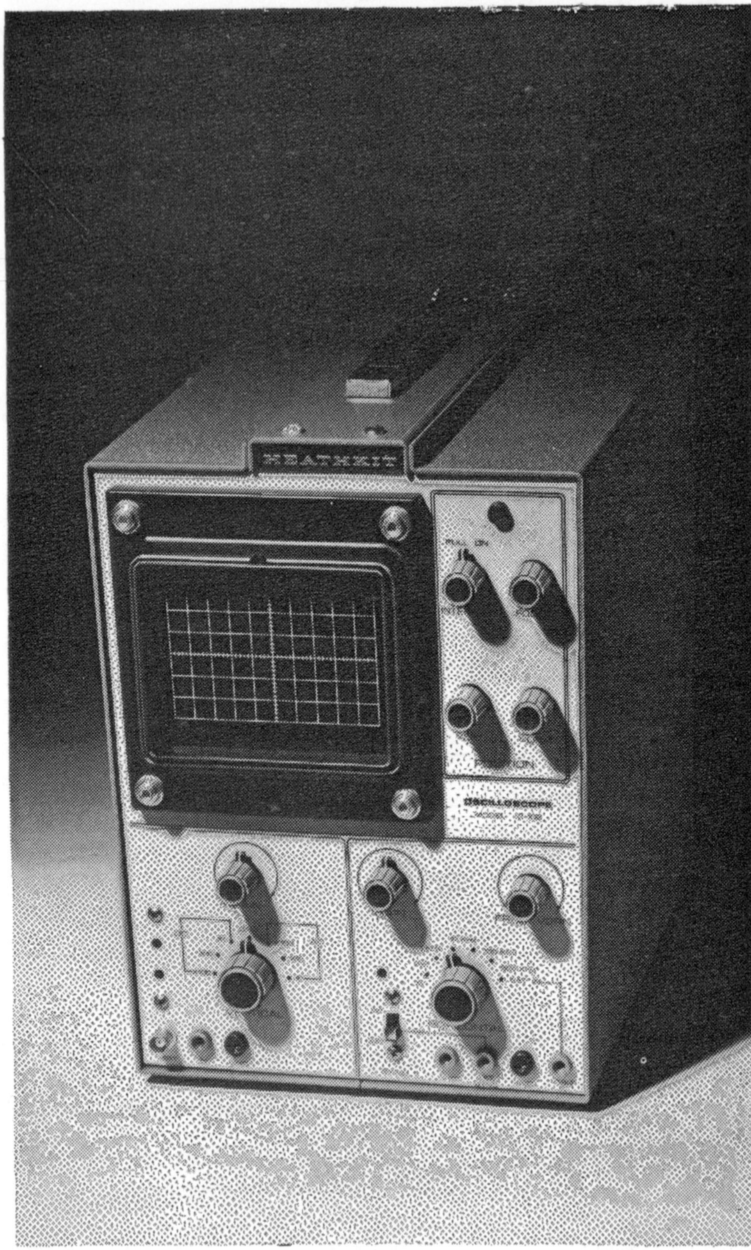

HEATH IO-102 OSCILLOSCOPE

Specifications—Heath IO-4510 Oscilloscope

VERTICAL

Input Impedance	1 MΩ shunted by 38 pF.
Maximum Input Voltage	400 volts peak combined AC and DC.
Sensitivity	1mV to 5μV/cm in a 1-2-5 sequence, accurate to within 3%. Uncalibrated, continuously variable between steps, minimum sensitivity of approximately 15 V/cm.
Frequency Response	DC to 15 MHz (-3 dB).
Rise Time	24 ns.
Overshoot	Less than 3%.
Vertical Windows	2 minimum.
Vertical Modes	Channel Y1, channel Y2, Y1 and Y2 chopped, channels Y1 and Y2 alternately. Modes are selected by the channel Y1 position, channel Y2 position, and time base switch.

HORIZONTAL

Modes	Sweep, X+Y and external.
Time Base	.2S to .1 μS/cm in 22 calibrated steps, 1-2-5 sequence; accurate to within 3%. Uncalibrated, continuously variable between ranges, maximum of approximately 500 mS/cm.
X5 Sweep Magnifier	Accurate to within 5%
Input Impedance	Approxiamtely 200 kΩ.
Input Sensitivity	0.4V/cm (approximately). No attenuator; negative signal causes trace to move to right.
X-Y Sensitivity	1 mV to 5 V/cm in a 1-2-5 sequence.
X-Y Phase Shift	Less than 3° at 100 kHz.

TRIGGER

Delay Line	Allows display of at least 20 nS of the pretriggered waveform.
Bandwidth (normal mode at 1 cm deflection):	
AC	20 Hz – 30 MHz, typically 20 Hz – 45 MHz.
ACF	15 kHz – 30 MHz, typically 15 kHz – 45 MHz.
DC	DC – 30 MHz, typically DC – 45 MHz.
Bandwidth (automatic mode at 1 cm deflection)	DC – 10 MHz, typically DC – 25 MHz.
Sensitivity	Less than .5 cm.
External Input Sensitivity	Less than .5V, typically .1V.
External Input Impedance	1M shunted by approximately 30 pF.

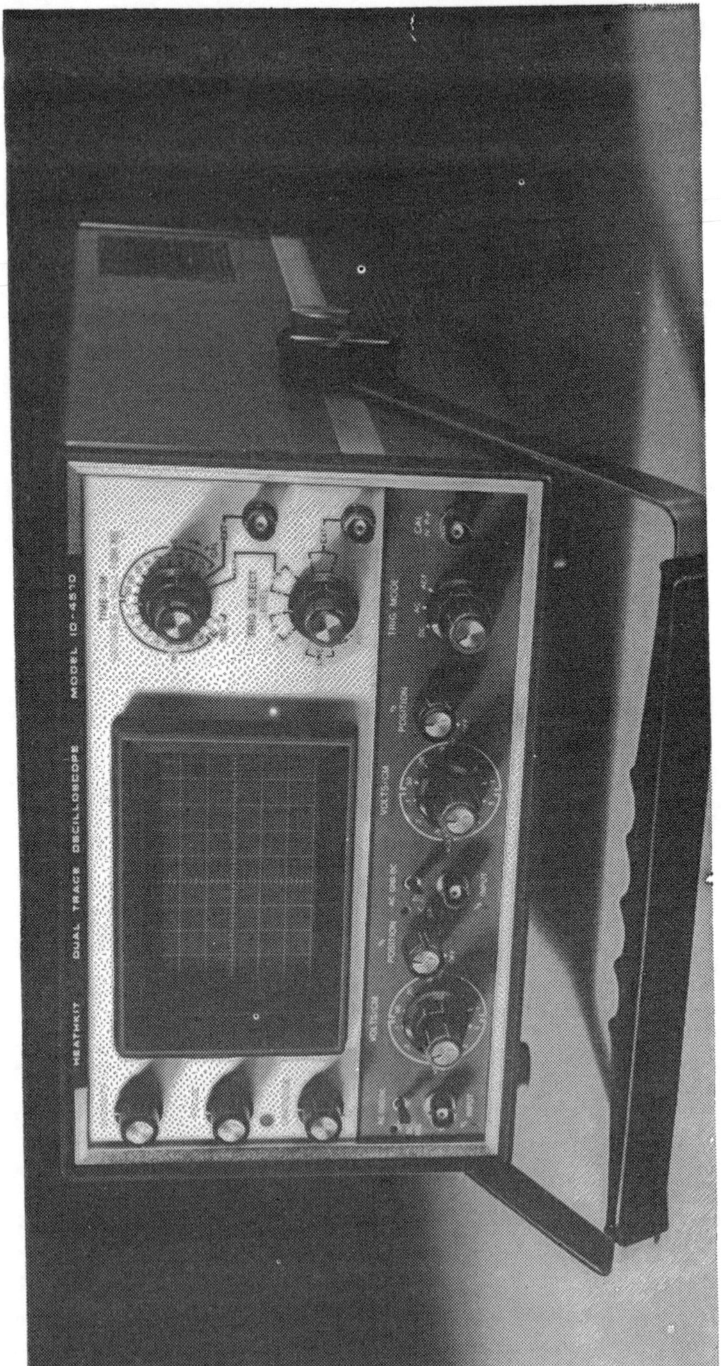

HEATH IO 4510 OSCILLOSCOPE

GENERAL

CRT Acceleration Potential . . .	4000V regulated.
CRT Type	6 X 10 cm Spiral PDA (Post Deflection Acceleration).
CRT Phosphor	P31.
Power Supplies	Fully regulated.
AC Line Switch	Allows operation from a 100-140 VAC or 200-280 VAC power source.
Power Requirements	100-140 VAC or 200-280 VAC, 50/60 Hz, 75 watts.
Input Connectors	Coaxial BNC.
Graticule Illumination	Adjustable from front panel.
Dimensions:	
Height	17.7 cm (6.937 in.).
Length (handle extended) .	54.5 cm (21.5 in.).
Length (handle folded) . .	49 cm (19.25 in.).
Width	32.5 cm (12.875 in.).
Weight	11 kg (24 lbs.).
Operating Temperature	10°C to 40°C.

Specifications—Telequipment CT71 Curve Tracer

COLLECTOR SUPPLY

Voltage Range—Peak voltage continuously variable from 0 to 1 kV, selected by horizontal VOLTS/DIV switch. Polarity is selectable, either positive or negative. The collector voltage repetition rate is twice the frequency or DC, selectable.

Peak Current—2 A; the peak power settings are 0.1, 0.5, 2, and 10 watts. Maximum power available is 15 watts.

Collector Series Resistances—Selectable: 0 Ω, 2.5 Ω, 10 Ω, 65 Ω, 250 Ω, 1 kΩ, 6.5 kΩ, 25 kΩ, 85 kΩ, 500 kΩ, and 1.7 MΩ, all within 5%.

BASE STEP GENERATOR

Current Range—0.2 μA/step to 20 mA/step in 16 steps (1-2-5 sequence).

Voltage Range—0.1 V/step to 2 V/step in 5 steps (1-2-5 sequence). Two positions are also available on the STEP AMPLITUDE switch to either open-circuit the base, allowing it to float, or short circuit the base to the emitter.

Steps/Offset—The steps are adjustable from 0 to 10 steps, selectable either positive or negative depending upon polarity switch setting. A continuously variable offset with a ± 1 step range is provided. Steps and offset are available on collector current ranges greater than 10 μA/div.

VERTICAL AMPLIFIER

Collector Current Range—Provides collector current from 5 nA/div to 0.2 A/div in 24 steps (1-2-5 sequence).

Horizontal Amplifier

Collector Voltage Range—Selectable collector or base voltage from 0.1 V/div to 100 V/div in 10 steps (1-2-5 sequence).

Cathode-Ray Tube—5½-inch CRT with a 10 x 10-cm viewing area. 2.5-kV accelerating potential with P31 phosphor. A front-panel control varies the graticule illumination intensity.

Power Requirements—Voltage settings are 100 V to 125 V in 5-V steps; 200 V in 10-V steps; 48 Hz to 63 Hz line frequency, 37 VA.

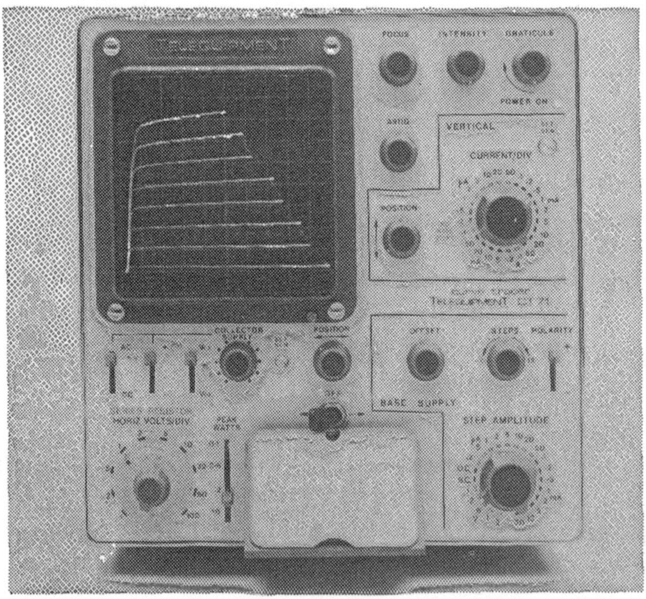

Dimensions and Weights		
Height	9.6 in	24.5 cm
Width	9.3 in	23.5 cm
Depth	19.0 in	48.3 cm
Net Weight	25.0 lb	11.7 kg

Included Accessories—Two test fixtures.

TELEQUIPMENT CT71 CURVE TRACER

Specifications—Heath IT-1121 Curve Tracer

Sweep Voltage Ranges	0-40 volts at 1 ampere maximum. 0-200 volts at 200 milliamperes maximum.
Sweep Voltage Sampling	.1, .2, .5, 1, 2, 5, 10, 20, and 50 volts/division ±3%.
Sweep Current Sampling	.5, 1, 2, 5, 10, 20, 50, 100, and 200 milliamperes/division ±3%.
Sweep Dissipation Resistors	0, 10, 50, 100, 500, 1000, 5000, 10 k, 50 k, 100 k, 500 k, 1 M ±10%.
Step Currents Available	.002, .005, .01, .02, .05, .1, .2, .5, 1, 2, 5, and 10 milliamperes/step, ±3%, ±250 nanoamperes offset current maximum.
Step Voltages Available	.05, .1, .2, .5, and 1 volt/step, ±3%, ±5mV maximum offset voltage.
Polarity Available	PNP and NPN (P Channel – N Channel).
Calibration Source	9 volts ±2% in 1 volt steps.
Oscilloscope Requirements	Vertical sensitivity of 1 volt/cm. Horizontal sensitivity of 0.5 volt/cm. Bandwidth to 20 kHz or greater. (DC-coupled oscilloscope is recommended.)
Operating Temperature Range	10°C to 40°C. Temperature variation, referenced at 25°C, will have a maximum effect of ±1% on all other specifications.
Line Voltage	110 to 130 or 220 to 260 VAC.
Dimensions	11-1/4" W x 10" D x 4-1/2" H.
Weight	Approximately 8-1/2 lbs.

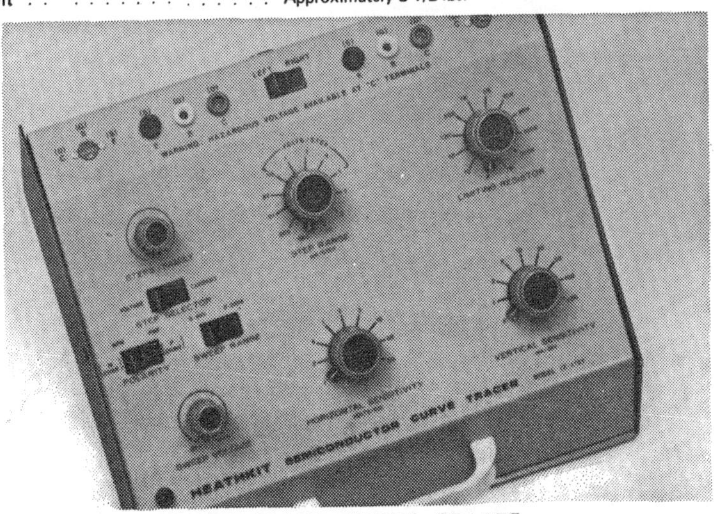

HEATH IT-1121 CURVE TRACER

Specifications—Heath PKW-101 Probe

Input Resistance	10 megohm.
Oscilloscope Input Capacitance Range .	Up to 40 pF.
Connector	BNC.
Maximum Voltage	600 Vdc or 600 Vac peak-to-peak.
Attenuation	10:1 (with oscilloscope having 1 MΩ input resistance).
Attenuation Accuracy	±3%.
Frequency Range	DC to 60 MHz.

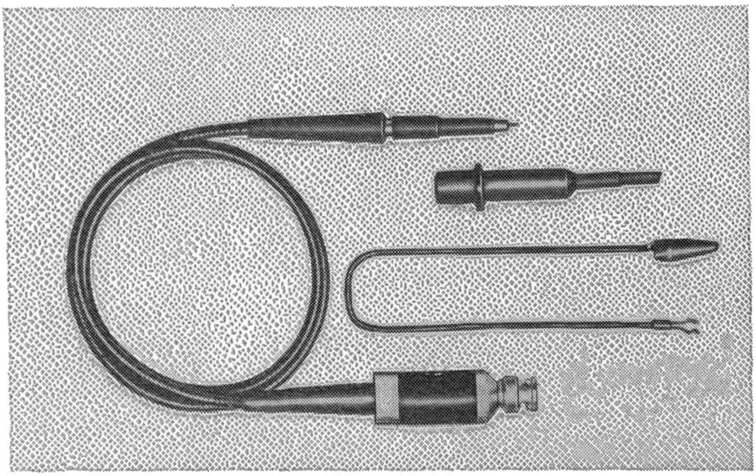

HEATH PKW-101 PROBE

Index

Index

U

V